# Swirling Currents

## Controversy, Compromise, and Dynamic Coastal Change

Sandy Macfarlane

Riverhaven Books

www.RiverhavenBooks.com

*Swirling Currents: Controversy, Compromise, and Dynamic Coastal Change* is a compilation of materials read and experience gained over the course of many years. Care has been taken to credit research and ideas.

Published in the United States by Riverhaven Books,
www.RiverhavenBooks.com

ISBN : 978-1-937588-83-0

Printed in the United States of America
by Country Press, Lakeville, Massachusetts

Designed and Edited by Stephanie Lynn Blackman
Whitman, MA

# Dedication

For Charles "Pete" Cole, Professor Emeritus of Fisheries Biology at UMass, Amherst, whose dedication to the success of his students and ability to think outside the box gave me the opportunity beyond the traditional approach to find my niche.

And for Fred Taylor, Professor at Antioch New England University whose positive teaching style helped me to find my voice and taught me the value of praise and encouragement as a life lesson.

# Table of Contents

# Table of Maps

# Acknowledgments

This book could not have been written without the assistance and support of many people with whom I have interacted professionally and personally. To those I have not mentioned by name or have inadvertently missed, please accept my apology and heartfelt gratitude.

A very special thank you goes to Nancy Cohen who volunteered to edit the first major manuscript draft in its entirety. Her skill, astute observations and suggestions were an invaluable contribution to the final document.

To Kim Tellert and Bev Wood, a special thanks for their thoughtful comments on the manuscript, artistic flair and ceaseless moral support.

Thanks go to authors Michael Barkham, Jeffrey Bolster, Eric Dolin, Brian Fagan, Mark Ferguson, Matthew McKenzie, and Stephen Palumbi for their generosity in providing clarifications and conversations for specific questions. I am also indebted to Jim Coogan, Dana Eldridge, Nancy Lord, Lee Martin, Dinty Moore, Roland Russell, Sherry Simpson and Fred Taylor, for their valuable feedback and suggestions throughout the process.

To Bonnie Snow, keeper of Orleans' collective memory, for sharing her incredible knowledge of the town and its people. And to the Orleans Historical Society for assistance throughout the entire project.

Thanks go to reference staff and librarians Tavi Prugno at Snow Library in Orleans, Julie Rose as the NOAA Milford laboratory, Mary Sicchio, Cape Cod Community College Nickerson Room, and Colleen Field, Librarian at the Center for Newfoundland Studies. Also a big thanks to Bill Burke, Cape Cod National Seashore Historian, and people at the Provincetown History Preservation Project, Chatham Historical Society

Museum, Bourne Historical Society Museum, and Wellfleet Historical Society.

Gratitude goes to Linda Coombs, Jessie Little Doe Baird, Earl Mills and Nancy Eldridge and the staff at the Mashpee Wampanoag Museum for their kindness and generosity in sharing their history and culture.

For technical review, I am immensely grateful to Dr. Judy McDowell, Woods Hole Sea Grant Director, Phil Coates, Director of the Massachusetts Division of Marine Fisheries (retired), Pat Fiorello, Communications Director of the Atlantic States Marine Fisheries Commission (retired), Dr. Anne Giblin, Ecosystems Center Marine Biological Laboratory, Woods Hole, Dr. Matthew McKenzie, University of Connecticut, and Geri Lambert.

To folks in writing groups, Hugh Blair-Smith, Tamsen George, Melanie Merriman and Barbara Sillery; and also to Marj Burgard, Sandy Emler, Marguerite Lentz, Molly Lofgren, and Don Teare who all helped make this work better, I am immensely grateful.

To the Karr family, Diane, Ron and Emily, for their suggestions, enthusiastic support and valuable assistance and who recognized the broad-based potential for this book.

The staff at the Center for Coastal Studies were highly supportive and generous in their time and expertise throughout the genesis of this book. To all I am grateful but most especially to founders Graham Giese and Charles "Stormy" Mayo, Executive Director Rich Delaney, Mark Borelli, Amy Costa, Pat Hughes, Scott Landry, Laura Ludwig, Catherine Macort, Deb Magee, Owen Nichols, and Lisa Sette. I am also grateful to the staff at the Wellfleet Bay Wildlife Sanctuary, particularly Bob Prescott, Director, Melissa Lowe Cestaro and Jeanette Kerr.

Many other people offered valuable suggestions, insights and moral support. To all, I am extremely grateful: Olga Amaral, Cheryl Berman, Brenda Boleyn, Janet Brown, Joan Caldwell, Karin Delaney, Susan Devogel, Gef Flimlin, David Garner, David Groesbeck, Brenda and Tony Halter, Patti Holsclaw, Jane and Dave Hussey, Rick Karney, Dale Leavitt, Dot Leonard, Henry Lind, Dorothy Jean McCoubrey, M. Patricia Morse, Myke and John Moss, Diane Murphy, BJ Newmier, Carol Nuss, Kathy

O'Donnell, Bob Rheault, Kathy Rhodes, Brian Sharp, David and Joan Sheehan, Leslie Sturmer, David Wright, Marcia Wright and friends far and wide in the New England Estuarine Research Society, National Shellfisheries Association, East Coast Shellfish Growers Association and Coast and Estuarine Research Federation.

To Steve and Kerry Willard for the many wonderful hours aboard the *Swamp Yankee* and encouragement for this project.

To Bob Haskell, editor and Stephanie Blackman, publisher, for making this book a reality.

To all the photographers who so generously gave permission to use their images for this book.

To Herb Heidt and Eliza McClennen for once again producing such wonderfully detailed maps where every request was met with enthusiastic execution.

Lastly, to the many researchers and regulators, fishermen and farmers, volunteers, and people interested and engaged in any part of the marine world or the land uses that impact the marine world with whom I have had the pleasure to know and work with, thank you very much.

# Introduction

During one remarkable day early in 2017, more than two hundred of the highly endangered North Atlantic right whales, nearly half of the entire world's population, entered Cape Cod Bay to feed on the annual bloom of copepods, tiny shrimp-like animals. News spread fast, detailed in national stories and news programs. Throngs of people traveled to beaches for a glimpse of the elusive giants or their distinctive, V-shaped spouts. The presence of those whales and their habitat represent one animal whose protection leads to biological, social, political, and economic ramifications that impact shipping and fishing industries and, in turn, consumers.

The event highlighted how an event, perceived as local, is often part of a much larger picture. Cape Cod has a front-row seat to observe several global marine conflicts swirling around its waters.

Cape Cod's unique geography, an arm of sand and glacial till sticking more than forty miles into the Atlantic Ocean off the coast of Massachusetts, make it a physical boundary between cold water species to the north and warm water ones to the south. Called a biogeographic boundary, it joins the Bay of Fundy to the north and Cape Hatteras to the south as similar dividing lines. Cape Cod's position between the other two is an important location to observe many of the rapid changes in species distribution and seaside landscapes occurring all along the coast. Examining what is transpiring in Cape Cod waters is instructive for areas both north and south of the Cape to give a broader perspective about what is happening elsewhere.

Many of the issues being discussed are of a vast scale, and people may wonder what the relevance is on a personal basis. When the issue is brought to a local level, it makes it more real and, in a way, easier to understand. It might also spur individuals to action. That thought goes to the heart of the slogan: "Think globally, act locally."

For concrete examples of why Cape Cod is an important example of the larger picture of global marine conflicts, consider the following. More than half of the world's population of the highly endangered North Atlantic right whale routinely enters Cape Cod Bay. Additionally, Cape waters support one of the largest populations of humpback whales in the world. Annual aerial surveys conducted by the Center for Coastal Studies in Provincetown photograph and track individual whales, and the center maintains one of the most detailed long-term studies of a baleen whale population.

Between 2012 and 2017, over two hundred dolphins and fifteen hundred cold-stunned tropical sea turtles landed high and dry on Cape Cod beaches. These events cemented the Cape's distinction as one of the top three places in the world for where marine animals are stranded.

The highest eastern U.S. confluence of harbor and gray seals are now seen in Cape and Nantucket Island waters. Currently, thousands of pups are born each year, and thousands of more seals patrol the waters surrounding Cape Cod searching for fish and luring great white sharks to the area, terrifying beachgoers at one of the Cape's major tourist attractions, and sending economic shock waves through communities. Fishermen watched their industry collapse as invading fleets of foreign factory ships scoured prolific fishing banks from Newfoundland to Georges Bank. Regulations increased, numbers of fish-eating seals mushroomed, and New England fisherman felt the noose tighten around their traditional way of life.

The wild-harvest shellfish industry plummeted during the twentieth century. Oysters, once a prolific species, were nearly ubiquitous in Cape waters, and tremendous amounts of their shells were found in deep shell middens left by Native Americans. Now, oysters are no longer found in the estuaries unless they are cultured, and shellfish aquaculture on all of our coasts is one of the bright stars, an expanding industry that provides jobs, food, and environmental services to the public.

The Gulf of Maine, including Stellwagen Bank National Marine Sanctuary, the major feeding area of the whales, is heating up faster than just about any other place on the planet with unpredictable consequences. The Gulf is also expected to experience a greater sea level rise during the coming decades. Cape Cod is demonstrably exhibiting geology in action and is disappearing through erosion.

Finally, the Gulf Stream that drives the global conveyor belt of oceanic currents which affects climate around the world is slowing down with unpredictable consequences. That dramatic shift threatens animals and industries dependent on the seas.

The public reads the headlines and asks, "What is going on?" Each of the headlines has a backstory of intertwining factors. The example of the seal/shark/fish/fisherman conflict may have risen to the top of the headlines, but anything to do with fisheries is complicated. We see "Product of Iceland" tags on cod in the supermarket and wonder why, in the place named for the fish and in the state that for centuries has had a likeness of the "Sacred Cod" hanging in the State House, Massachusetts can no longer supply enough cod to meet the demand. While cod landings have plummeted, lobster catches have increased. Cod eat lobster eggs, so without the cod as predator, lobster landings are high. But lobsters have moved out of Long Island Sound because of warming waters while Massachusetts and Maine lobster fishermen worry about their future. We are left wondering what the price of lobster will be if we have to import them from colder waters. The web goes in many directions simultaneously — seals, sharks, cod, lobsters, fishermen, climate change and consumers.

The problems have not escaped the notice of the international community. A report issued by the United Nations Food and Agriculture Organization in 2010 stated, "80 percent of the world's fish stocks are reported as fully exploited or overexploited requiring effective and precautionary management." The Nature Conservancy published a report in 2009 stating that in most bays around the world, shellfish reefs contain less than 10 percent of their original abundance, and 85 percent of oyster reefs are completely lost and considered functionally extinct in many parts of the world. After protective measures were enacted, some species of marine mammals have made a comeback while others that had been hunted to near extinction have not. And now, climate change is altering the planet in ways that we have never experienced.

How did we get to this point? The past provides some clues related to our complicated relationship with the sea, but it is not the whole story. Focusing on a specific area such as Cape Cod as an example allows us to somewhat comprehend our world as it is today. Although not unique, Cape Cod represents a highly visible microcosm of problems affecting communities along all of our coasts. With an eye trained on the larger picture, lessons learned may be applicable to other regions as we look to the future.

None of the changes take place in a vacuum. Species don't live in a vacuum, and land changes don't occur that way either. Species and habitats are intertwined, and all of it is connected to us. Perhaps our awareness can be roused to the myriad linkages and our complex relationship to the sea through stories behind the headlines. The marine industries — fishing, whaling, and aquaculture; the target animals of those industries — cod, whales, and oysters; the specific environments — open ocean, estuaries, and rivers; and the locations — North Atlantic, Gulf of Maine, and Cape Cod — are all interconnected. The resources, environments or habitats, and people are woven together into the drama taking place in the seas.

Conflicts with the sea are not new. Humans have always had a strong and often volatile relationship with the oceans, although the issues may be more complex now. Tremendous population growth, rising seas, and loss of marine resources and habitat that are occurring everywhere amplify all aspects.

I knew how fortunate I was to live my dream of a career on the water. I was appointed as the shellfish biologist and conservation administrator for the Cape Cod town of Orleans where, throughout my entire life there, I witnessed tremendous changes, both physical and cultural. I spent every summer of my childhood in a tiny cottage near the Orleans-Eastham town line, a gift to my family to escape the industrial area where we lived. At an early age, I knew the difference between sandy beaches with clean water for swimming and seaweed-choked, slimy, smelly areas that others called beaches where I grew up. Like most little kids, I reveled in the sheer joy of being in and around the water, first at beaches where the water was calm and later in the ocean surf. But my curiosity was awakened at an early age to the creatures that lived in those waters. I watched life in action in tide pools left by the ebbing tide on the expansive sandflats of Cape Cod Bay — the periwinkles and fiddler crabs and minnows and starfish. I saw the remains of life on the flats, the shells of scallops and clams, the paper-thin jingle shells, purple tinged thick quahaug shells and barefoot-cutting razor sharp clam shells, moon snails, and the occasional sand dollar. They fascinated me. Back home, winter television specials produced by Jacques Cousteau further stimulated my curiosity. Summer and winter experiences merged and I knew that I wanted to call Orleans home as soon as possible. I moved there directly after finishing college.

I watched as tourism led the economic way. As tourism grew, the Cape population also increased. Land development for houses, businesses, and municipal infrastructure increased and formed the foundation and building blocks of a tourist-based economy catering to a constantly rising influx of seasonal people and year-round residents. With that increase, serious and contentious issues emerged to threaten Cape Cod in the twenty-first century. When tourism replaced commercial fishing as the primary industry, fishermen who hung on became highly regulated as to what they could harvest, altering their lives.

Our resource management decisions have generally been based on a species-by-species approach. As one species dwindled, we devised methods to limit harvest of that species. But we have done so by looking at the life cycle of the species alone and not necessarily within the context of its interactions in its own environment. We now accept that the exploitation of one species may have a devastating cascading effect on other species. We do not know if it will happen, how it will happen, or what it will cost. But species-by-species management is an inadequate approach for long-term sustainability. Other methods are needed and are being evaluated now.

Resource management is much more than just biology. Marine resource management deals with biology of course, but also with people — social, political and economic factors as well. Human wants and needs are often at odds with each other, but they may also be at odds with the requirements of the resources we are trying to manage, making resource management a difficult juggling act. Practitioners try to predict the future based partly on the past and partly on incomplete information about the present. Decisions are made within the context of an ever-changing and sometimes dizzying array of possible outcomes while investigations continue. My love of the marine environment and desire to work toward protecting it, and the creatures that inhabit it for the present and for future generations, had been nurtured as a child when I spent every summer in that tiny cottage across the street from the Town Cove, part of the Nauset estuary that I would come to know so well as an adult.

As a resource enforcement officer I witnessed the power of the legal and court systems firsthand and understood the importance of the language of regulation. I recognized the difficulty for municipal conservation commissions, volunteers who are charged with the responsibility to approve or deny permission for projects. It was an overwhelming, difficult, and potentially wrenching responsibility for commissioners to tell friends and neighbors what they could and could not do with private property. And in Massachusetts, private property rights are paramount – culturally, historically, and, of course, economically.

Added to the human factor in resource management are natural changes. Steady encroachment of the sea because of rising sea levels, exacerbated by winter storms, challenges our life on land. Those same storms can affect species that live in the water where all may be different after one storm. Fishermen have traditionally harvested what was readily available, knowing through experience that a steady supply was not always possible. That uncertainty prompted them to harvest today and worry about tomorrow's harvest, tomorrow. That attitude, mixed with other factors, prompted managers to regulate harvests.

I think about the constant parade of new residents who see things the way they are when they move to a new area. Many of them think that what they see is the way it has always been and the way it should always be. What they see and experience form their personal baseline from which they gauge the world around them. They have no point of reference to differentiate the past as distinct from the present or to visualize alterations that will take place in the future.

I was at Nauset Beach one day in March 2018. Winter Storm Riley had paid a fierce visit to the region, and the beach had been pummeled. Water had lapped at the base of the gazebo at the edge of the parking lot. Going back to the time of town bands, that gazebo had been there for many decades, and had been the site of evening concerts for throngs of people. The waves had also hit the foundation of Liam's, the food concession that had been there since the 1950s when it was known as Philbrick's Snack Shack, famous for fried clams and onion rings. The gazebo was moved to a safe location, and Liam's was demolished by the end of March. People at the beach lamented the loss, many posting online about how sad it was. Sad, yes. Inevitable, yes. The end of an era, yes. And proof to dispel any illusion to the contrary that when Mother Nature flexes her muscles, we are definitely not in control and all we can do is observe in awe.

Our relationship with the sea loses none of its complexity as our understanding of the water and its inhabitants evolves. Part of that knowledge comes from examining the whole habitat as well as its parts to discover subtleties as well as the obvious. But many of our problems with the sea around us are of our own making, and we need to come to terms with the effect of our activities. It has taken us centuries to acknowledge and embrace the notion that humans are only a part of the ecosystem, incorporating the concept of interconnectedness in our everyday world. We need to understand the past and blend that knowledge into policies for the future in order to move forward.

As we move forward, we will need interdisciplinary and interinstitutional collaborations, government and institutional projects, as well as public and private liaisons. With partnerships and perseverance, perhaps we can deal with problems in greater harmony with the natural world and her rhythms than we have done previously. By examining our historical and current relationship with marine resources — marine mammals and endangered species, fisheries and shellfish, and habitats — and interpreting biological, social, political and economic factors, we can better understand our relationship with them. While cumbersome and highly complex, it may be the only way to resolve extremely difficult problems. There are no easy solutions and there is still so much to learn. But there are some hopeful signs to help point a way for the future. Many people are hopeful as the efforts unfold.

# Chapter 1
# Ground Zero

"Out of the water! Get out of the water!" The alarm in the lifeguard's voice, bellowing through a megaphone, was unmistakable. "Everyone out! Now!"

Lifeguards urgently forced people out of the ocean. Hundreds of people who had been body surfing, diving under waves, splashing and having a good time on a beautiful summer day in 2017 were forced to abandon the surf at Nauset Beach, in Orleans, a perennial favorite among Cape Cod's bountiful beaches. Parents scrambled to get their kids safely onto the beach. Teenagers without parental supervision defied the order for as long as possible. The lifeguards yelled, whistled, and gestured to get people out of the water — fast. A great white shark had been spotted just beyond the breakers in the shallow water merely thirty yards offshore, less than half a football field. Soon everyone knew the reason for the order. Only a few minutes earlier, beachgoers had oohed and ahhed at the close-up view of adorable seals that poked their heads out of the water to catch a breath. A large information kiosk at the beginning of the boardwalk to the beach warned of the possibility of sharks, but the reality really got the adrenaline pumping.

The beach was cleared again a few days later for another shark sighting. This time the water roiled. Suddenly a great white shark shot out of the water clutching a seal within the multiple rows of its razor-sharp, serrated-edged teeth. The water turned blood red in full view of the tourists on the beach. Many expressed horror tinged with awe at the raw spectacle of the scene while others lamented that their vacation was being ruined by the constant threat of sharks.

The events were added to the series of headlines about great white sharks in Massachusetts waters. The previous September, a shark took a bite out of a kayak in the calmer waters off Plymouth, on the western side of Cape Cod Bay. The news was broadcast on local and national channels. The two kayakers were traumatized but safe. A year before that, a man was attacked by a shark off a Truro beach. He survived, but the scars to him and the communities remain.

At the docks, another drama unfolded as angry fishermen, expletives loudly filling the air, griped about competing with seals that steal fish from their hooks or nets. Stories about this competition for fish and conflicts surrounding seals/sharks/fish and fishermen frequently appear in newspapers.

Thousands of seals now patrol the waters off Cape Cod beaches in search of fish while sharks pursue their favorite food, the seals, and, in the process, terrify humans. Great white sharks were a rarity a decade or so ago but were becoming an all-too-familiar occurrence by 2017, appearing along Cape shores in surprising and threatening numbers.

Why the increase? It's simply because gray seals have increased as well due to protective measures in place. The spike in great white sharks, and a mushrooming population of seals have created a different reality from several decades ago. Lifeguards, harbormasters, and first responders have their hands full searching for telltale fins. Those fins are striking fear, causing a decline in beach revenues and creating a deep sense of unease about venturing into the ocean.

The seal/shark/fish saga is not the only conflict in the news. Cape Cod has become a focal point for many marine conflicts, each producing attention-grabbing headlines.

The public reads the headlines and wonders why all this is happening. Exploitation is the simple answer. We have taken resources from the sea perhaps believing, or wanting to believe, that these natural resources are capable of self-restoration and will always be there to exploit as before, rarely considering them to be gifts from the sea. Basically, we never truly learned to live in harmony with the world around us. We want it all.

This was not always the case. For the Wampanoag tribe, the first inhabitants, including the Nausets, a segment of the tribe on the eastern third of Cape Cod, the bountiful and richly diverse waters they relied on defined much of their lives. It was a part of their culture to understand the life cycle of any animal, fish, or bird they took for food, and to recognize seasonal changes that provided availability and variety. They considered the sea's resources they depended on as gifts from the Creator, to be received with thanks and gratitude. Although trade was an important aspect, those gifts were not treated as commodities for sale as we know the term.

Conversely, the European way of life that required resources for survival also emphasized the economic benefit to harvest those resources. Generations of tradition fostered an attitude to take whatever they wanted whenever they wanted in any amount. And that often meant high profits. That attitude has been passed down to us, and it has created problems with the way humans relate to the sea today regarding marine habitats and the mammals, endangered species, fisheries, and shellfish in the sea.

Fort Hill in Eastham, Massachusetts, part of the Cape Cod National Seashore, is a perfect place to think about the marine world of the past, present, and future. Thanks to some bold political visionaries, the hill is a public place, one that was almost not included in the park. When federal officials and local representatives flew over the site during a helicopter reconnaissance flight to plan for park boundaries in the 1950s, the hill was covered with subdivision stakes. They concurred that the exquisite property had to be part of the public trust so everyone could enjoy the magnificent vista, not merely the few who could afford to own a piece of this paradise. It is now one of the park's most popular vantage points, often thought of as the crown jewel of the Cape Cod National Seashore awaiting discovery by many of the five million visitors each year.

Surrounded by rolling fields and walking trails that wind through a protected and fragile ecology, the parking area at the summit affords a panoramic view of where water and land merge. Geography, history, culture, economy, and ecology converge at this small, remote, watery locale; and the sea continues to rule. Marsh hawks glide on unseen wind currents and hover momentarily before diving toward the fields below in pursuit of food. The salt marsh, an expansive blanket, changes color depending on the season. The green of summer mixes with gold-tipped grass in late summer and fall and then turns a tawny brown by winter. Ribbons of blue meander through the marsh as creeks carry the tidal waters to the headwaters of the Nauset estuary on the flood or back to the ocean as the tide ebbs. Exposed intertidal bars dot the creeks, home to clams and quahaugs beneath the surface while blue-black mussels visibly border a few exposed flats.

East of the marsh, the white-sand barrier beach, appearing like a watchful sentry, guards the estuary from the fury of the open ocean. The beach is interspersed with greenish-brown beach grass that anchors the dunes in place. All is calm on this day. But when a storm comes, stark change happens overnight. The storm rips away both the beach grass on the dune and the fifteen-foot rhizomes below the surface, becoming detritus in the marsh and along the shore. Sand washes into the bay behind the dunes, and the barrier beach as a whole unit physically takes a hike westward.

Farther east, beyond the pencil-thin line of protective sand and dunes, lies the vast Atlantic Ocean in all its moods.

Surrounded by ocean, dappled with numerous bays, coves, harbors, estuaries, beaches, and fresh ponds, water defines how Cape people survived daily challenges and still do. The sea that constantly encroaches, characterizes a peculiar geology and geography that helped a hardscrabble, resilient people evolve on Cape Cod. They lived and worked in a harsh landscape where the forces of the sea influenced their lives and economic development. They still live in a sort of schizophrenic manner — appreciating and communing with nature or attempting to assert dominance over it — sometimes simultaneously. As they do so, the Cape culture, like its dunes and inlets, always reinvents itself. But there are signs indicating that current conflicts threaten to unravel the land, water, and cultural threads of the Cape Cod tapestry and reweave them into a completely different and possibly unrecognizable pattern.

Visitors today, who revel in the opportunities that the sea provides, may not realize how the Cape became such an evocative place, how its maritime history was warp and weft in the context of the present weave or how the power of the ocean determines its destiny, or how the Cape is tied so intricately to global marine resource issues.

Once the ocean vista was rich with hundreds of ships sailing up and down the coast, necessary for the maritime industries that enhanced the region's economy. When I stand on Fort Hill and look beyond the beach to the Atlantic, I imagine those hundreds of sailing vessels, but I see only one lone vessel. And it is not under sail. It is motoring upon the sea with an internal combustion engine, a symbol of tremendous change. The sea enabled fishing, whaling, maritime commerce, and salt-making to become industries that were integral to the region's evolution and made Massachusetts an economic power.

Commercial fishing remains important but hardly resembles fishing of the past. In the distance I watch a few commercial fishing boats thread their way back to their moorings after successfully negotiating another trip through the treacherous Nauset Inlet. The harbor mapped by Samuel de Champlain, who dropped anchor inside the harbor in 1605, is not the same place now. It is so shallow that the men moor their boats in the channel near the inlet, making their daily chore of loading and unloading gear and their catch even more cumbersome than it was not that long ago.

While commercial fishing has declined precipitously, recreational fishing has steadily gained prominence as a multimillion-dollar industry in New England, with multiple ancillary businesses, as anglers search for bass, blues, or the elusive giant tuna. Weekends are especially busy at marinas and launching ramps everywhere. As soon as the first bass is seen, the frenzy begins, lasting until the bass leave the area in the fall and the boats are taken out of the water.

Catching whales is no longer a U.S. industry, but whale watching is. Passengers on whale watch boats learn that the survival of endangered and threatened species is in peril as they now compete with people for their very right to exist.

When out on the vast ocean, and out of sight of land, it is easy to imagine that the resources of its great depths are inexhaustible. We are slowly learning, however, that the ocean is indeed exhaustible and is not as full now as it was when Europeans first reached these shores.

Fort Hill evokes a serene calmness, but that placid veneer disguises swirling undercurrents that will be difficult to resolve in the twenty-first century. They are part of a web of interconnected circumstances, natural events, and human activities that result in alterations to the landscape — and to our perceptions. At this bucolic scene I see the obvious beauty, but that essence is veiled by representations of something else entirely. I see the parts, but I see the whole as well — the result of a forty-year career in conservation and resource management. I see signs of how we have impacted marine resources — the habitats and the creatures that depend on those habitats. And I recognize that Cape Cod and the region merely represent the conflicts that we, in general, have with the sea and its inhabitants.

But let's start at the beginning.

Chapter 2

# Glaciers: They Came, They Dumped, and They Left Cape Cod Behind

No one was there to hear the deafening thwack, or witness the twenty-five-story, city-block-sized chunk fall away from the ice wall into the lake, or marvel at the massive waves that fanned out from the impact. The ice sheet was melting. Water seeped into crevasses to form a slippery river beneath the enormous sheet, while gravity worked its magic to break off successive bergs of weakened ice that crumbled into the lake. With the calving of each block, the lake's water level rose, bulging at its seams. It wanted to drain.

After millennia, it did drain. The glacial lake that became Cape Cod Bay was hemmed in by the retreating main ice sheet to the north, by the South Channel Lobe to the east, and by high hills of debris left by the retreating Buzzards Bay lobe on the west and the Cape Cod Bay lobe on the south. Erosion wore down some of the debris over time to form at least three valleys that were just low enough to allow water to flow out of Glacial Lake Cape Cod.

When the lake drained, the rivers and lake bed were dry land. When the glaciers melted and retreated, the tremendous amount of water that had been trapped in the ice caused the sea level to rise, submerging the dry land, not unlike sea level rise now. The higher glacial deposits around the rim of the former lake began to take the now familiar shape, leaving behind Cape Cod Bay and Cape Cod.

Glaciers move. In the last glaciation period, whole mountains were covered and much of the continent was under an ice sheet. It was as if an unkempt bed with lumpy blankets was covered with a thick white coverlet that hid the ridges and depressions of the blankets. Mountains and valleys were invisible. In some places, the ice was over a mile thick, nearly four Empire State Buildings deep. A very slow unstoppable force was at work.

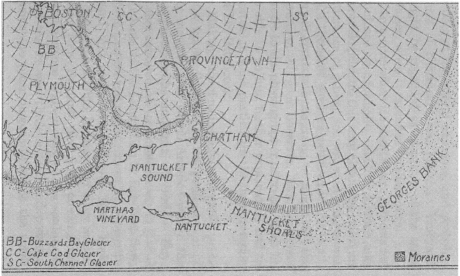

**Geological map indicating the three glacial lobes that formed Cape Cod: Buzzards Bay, Cape Cod Bay and South Channel. As the glacier retreated, the Cape Cod Bay lobe was hemmed in by the material left by the Cape Cod Bay lobe and the other two glacial lobes, forming Glacial Lake Cape Cod Bay. The lake eventually drained and as sea level rose, Cape Cod Bay was created. Cape Cod Commission**

Anyone who has watched an ice floe in a river knows that the tremendous energy and force of the ice will bulldoze whatever is in its path. Imagine the power of ice, higher than many mountains, moving as a single unit and what it would be capable of moving. As the ice sheet expanded, adding weight, its nearly imperceptible movement pushed loose material farther and farther south. The process took thousands of years, not decades. The time scale of glaciers and their movement is not in our experience any more than is the magnitude. When it stopped its march southward, it left behind a ridge, thousands of feet thick – the moraine. Geologist Robert Oldale uses the analogy of a bulldozer when he describes how the ridge was formed in his 1992 book *The Geologic Story*. The moraine has a shallow angle on the glacier side and a steep angle on the opposite side just as a bulldozer creates a ramp to the top of the pile and dumps the load that falls nearly vertically from the top. The moraine on the Cape indicated that ice stopped in the mid-Cape area, but Martha's Vineyard and Nantucket each have moraines too. After the moraine was formed on the Cape, another "cooling" period occurred whereby the ice sheet again expanded south to the islands and offshore to George's Bank where it finally stopped its southward expansion and began retreating northward.

At 305 feet, Bourne's Pine Hill is the highest point on the Cape that remains from the moraine. As the glaciers melted and retreated, all the sand, gravel, clay, and boulders were left behind. Torrents of water that we cannot begin to imagine were let loose in meltwater streams and braided rivers that flowed downhill to a sea level that was about three hundred feet lower than it is now. From the steep face of the moraine, the water deposited the glacial till and formed the outwash plain — debris washed out of the glacial ice and extended as far offshore as George's Bank.

The Outer Cape outwash plain slopes from east to west. A dramatic view of the material is the scarp face of Nauset Beach from Eastham to Truro. There, the cliff rises 120 feet or more above the beach in layers of sand and clay with boulders scattered throughout. Erosion has lowered the height of the scarp considerably. What it was when the glaciers actually retreated is supposition. However, between three hundred and six hundred feet of till lies beneath the ground before bedrock is reached, underscoring the Cape's fragility. At the lower end of that scale, that amount of flimsy material overlying bedrock is nearly as deep as the Pilgrim Monument in Provincetown stands above the ground surface, 350 feet above sea level. The western side of the outwash plain is visible in a slightly less dramatic scarp on the Cape Cod Bay side. One would also expect to see a moraine on this part of the Cape, but it does not exist. It eroded out to sea long ago.

Kettle holes and ponds were created when the multi-thousand-foot-thick ice sheet retreated. Partially or completely buried chunks of ice left behind were surrounded and insulated for centuries by sand, clay, and gravel washing out of the glacier. Eventually, they melted. The till surrounding each block collapsed to form a kettle hole. Some filled with fresh water and may be landlocked or they have an outlet creek that connects them to the sea. Those that connect to the sea may have become habitats for spawning fish like herring. Still others, called drowned kettle ponds, filled with salt water as sea level rose when the glaciers retreated and became salt ponds in estuaries where fresh water meets the sea. The Cape is left with over one thousand fresh water lakes, ponds, and creeks; a hundred or more salt ponds, rivers, and coves; and a mixture of both fresh and salt water that creates the fantastically productive estuaries with their outstanding habitats.

These processes — Glacial Lake Cape Cod, the moraines, outwash plains and kettle ponds — formed the basis of the evolving landforms that followed and continue to evolve. The processes define the coastline of southeastern Massachusetts — miles of sandy beaches, tidal flats, shallow near-shore feeding grounds, offshore banks and basins, dunes and barrier beaches, coastal banks and kettle ponds. It is a legacy of spectacular beauty and bounty recognized by people who arrived to explore, exploit, and enjoy what the land and water had to offer and still do. The accident of the location of the Cape, extending forty miles off the coast of Massachusetts, added another dimension to its story as the closest unobstructed mainland point to Europe and proximity to plentiful seas.

Estuaries with numerous bays and coves teem with life; fantastically productive salt marshes and eelgrass meadows offer food and protection to hundreds of species; streams lead to fresh water lakes and ponds; the sandy soil drains well — sometimes too well; bogs and fresh water wetlands and acres and acres of waterfront property line the shores of both fresh and salt water. Taken together, the variety of distinct habitats is one of the attributes of the place.

Geology seems to be a static science. Mountains stand in magnificent and massive splendor, but erosion occurs all the time. Water and wind wear them away, but the changes are subtle. Then something like a massive rock formation resembling a man's head falls off a mountain like it did in 2003, and suddenly the state symbol for New Hampshire is gone. At that point, we may acknowledge geologic forces at work.

On Cape Cod, a land of sand, geology is dynamic and happens before your eyes; erosion and deposition are facts of life. Summer visitors who never see the Cape in the winter are amazed at the differences year to year or even season to season. Stand on a bluff overlooking the Atlantic during a nor'easter, and the wind and water overwhelm the senses of sight, sound, touch, smell, and even taste. Or go to a beach at the bay-side elbow of the Cape during the winter, when ice piles into the curve as far as the eye can see, and the view of Cape Cod Bay is devoid of any blue water, appearing more like an Arctic scene than the Cape Cod of summer's pleasures. At that moment, the imagination may transport the visitor back to a world of glaciers when ice was all there was.

Geologists theorize that three prominent features of the Cape today were drainage points of the ancient Glacial Lake Cape Cod Bay. One formed at the margin of the Buzzards Bay Lobe and Cape Cod Bay Lobe. It eventually became the obvious location for the Cape Cod Canal. A second river formed in the middle that became Bass River, a river that now flows nearly all the way — but not quite — from Nantucket Sound clear through to Cape Cod Bay. A third location at the "elbow" of the Cape, became another river that formed at the confluence of the Cape Cod Bay Lobe and South Channel Lobe — Jeremiah's Gutter. Little attention is paid to Jeremiah's Gutter today, but it has an important and colorful history.

✦

The storm raged. A single small sail remained, as ineffectual as a handkerchief. The captain — even one like the infamous pirate Samuel "Black Sam" Ballamy — was powerless to steer away from the looming shore. Avoiding disaster was an impossible task as the howling winds pushed the desperate craft closer to land. It headed for the beach where breakers crashed one after another in a merciless onslaught. The crew was reduced to merely hanging on and praying, as the relentless hurricane-force winds and towering seas churned the ocean in all directions at once, tossing the boat up and down and side to side as if it was nothing but flotsam, while pelting rain blinded the men. They watched in terror as the "backside" beach of Cape Cod, and almost certain death, awaited. Death did claim 102 souls of the Whydah that fateful night of April 26, 1717, including Captain Bellamy. All the loot in the world could not save them as the ship ran aground on an offshore bar, five hundred feet from shore.

Within days, Massachusetts Bay Colony Governor Samuel Shute sent Captain Cyprian Southack to the site with specific orders to retrieve anything the Wellfleet residents had taken from the wreck and bring the items back to Boston. Southack stayed for a week, but his mission ultimately failed. Deadly waves, strong rip currents, and shifting shoals prevented him from boarding the wreck; and the people refused to

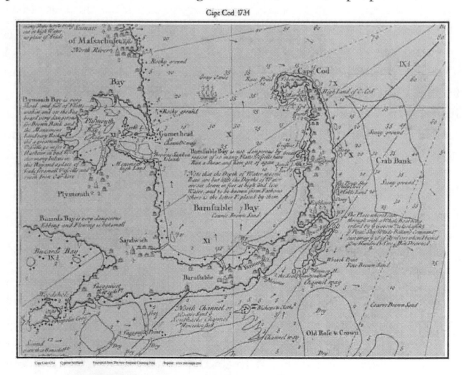

**Captain Cyprian Southack's 1734 map of Cape Cod clearly showing a river between Cape Cod Bay and Atlantic Ocean, later named Jeremiah's Gutter.**

relinquish the goods they had taken.

Captain Southack, a cartographer, had reached the Whydah by sailing his whaleboat through a strait in Eastham. He knew of the waterway when working on his New England Coastal Pilot, the most comprehensive nautical chart of its day, published in the 1720s. Southack's map depicts a passage from Cape Cod Bay to the ocean that left the area from Eastham to Provincetown as an island. He had confirmed Cape Cod's first canal — later named Jeremiah's Gutter.

Earlier, in 1659, a group of five adults and six children sailed from the Merrimack Valley toward Nantucket to begin a new colony. Hugging the coast, they arrived at the same passage and sailed through it to Nauset Harbor, wrote Eric Jay Dolin in his 2007 book, *Leviathan*. From there, they could travel through another inland passage from the harbor to Pleasant Bay and Nantucket Sound and, ultimately, to their destination. The inland passage is long gone. The marshes lining the creek gradually encroached farther and farther into the creek, filling it in to a narrower width, obliterating its usefulness as a passageway.

The river Southack found was well known to the Nausets, a segment of the Wampanoag tribe, who were living in the area. How deep it might have been in the distant past, no one knows. Thousands of years after the glacial lake drained, the bay's sands shifted, the marshes filled in, and the Town Cove decreased in depth, transforming and reconfiguring the stream.

It was known as Boat Meadow Creek on the Cape Cod Bay side. Farther inland, the narrow and shallow river flowing through the marsh, was known as Jeremiah's Gutter, named for Jeremiah Smith who owned the property. In 1717, Smith, who hailed from an early Cape family, deepened the passage that ran from Boat Meadow Creek in Cape Cod Bay to the Town Cove in the Nauset estuary. He created a canal, a shallow creek that could accommodate twenty-ton ships.

The canal was important during the American Revolution when British warships closed the Port of Boston and enforced a blockade. American ships that managed to slip out of the harbor traveled south along the Cape Cod shore and found Jeremiah's Gutter — a 100 to 150-foot wide channel to the Atlantic — but only at high tide. At low tide, exposed sand flats extend for over a mile into Cape Cod Bay. Later, enterprising U.S. captains used that trick again as a shortcut to the Atlantic during the War of 1812 when the British again blockaded Boston Harbor according to the Orleans Historical Society.

By 1820, a group called the Eastham-Orleans Canal Proprietors had tried to maintain the canal by charging ten cents per ton for passage. Unable to maintain the width and depth necessary to accommodate larger vessels, the venture failed after only three years. By the middle of the century, a new rail service reduced the need for ship transport, and the canal, which had fallen into disrepair, was no longer used.

The natural processes that shrank the canal's size over centuries finished the job, negating its commercial value and erasing it as a landmark. But, for at least 150 years, the passage begun by nature, restructured by Smith, and documented by Southack, made an alternative and safe route by boat to the Atlantic possible. Boats could avoid a day's travel along the treacherous and potentially deadly route around the back side of the Cape, with its shoals, bars, and breakers, explained Shawnie Kelley in her 2006 book *It Happened on Cape Cod*.

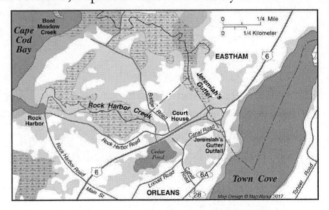

Jeremiah's Gutter represents something much larger than a two-foot-wide marsh creek now. It represents eighteen thousand years of natural and human forces in a constantly evolving landscape and an evolution of our relationship to the sea.

**Jeremiah's Gutter flowing from Cape Cod Bay to Town Cove in the Nauset estuary. Gray area indicates wetland and cranberry bogs. (Mapworks)**

Formed by retreating glaciers, it is now a focal point when looking ahead to effects of sea level rise. Once used as a watery highway from Cape

Cod Bay to Nantucket Sound, it may have also been a factor in species distribution, and it may contribute to that process again in the future. The changes over time are dramatic. Yet, unlike the outer shores of the Cape where change occurs in a matter of hours, the changes are subtle and mostly go unnoticed — until Mother Nature flexes her muscles. Then the changes are hard to ignore.

⁂

To the uninitiated motorist traveling on a multilane highway to the Outer Cape, a reduction in speed on a thirteen-mile, two-lane, no-passing section of the Mid-Cape Highway, U.S. Route 6, is often annoying. But a 25 mph speed sign comes as a shock. Another sign quickly follows — a circle of arrows indicating a traffic rotary or round-about circle lies ahead.

Approaching the rotary, motorists pass Cedar Pond on the right. The pond is connected to Cape Cod Bay by a marsh creek that flows into Rock Harbor. The creek is one of five that meander inland from Cape Cod Bay, surrounded by extensive salt marshes that line the curve of the elbow of Cape Cod Bay. Deemed exceptional natural resources to be protected, these twenty-six hundred acres of creeks and marshes were designated as the Inner Cape Cod Bay Area of Critical Environmental Concern in 1985. The state designation came after ecological concerns about construction and manmade alterations of such precious environments became important, and now projects needing state permits require greater oversight.

Less dramatic or decorated than other traffic circles on Cape Cod, the nondescript rotary straddles the town line between Orleans and Eastham. There is no sign carved in shrubbery or spelled out in flowers that says "Welcome to Orleans" or "Welcome to Eastham" as one might see elsewhere at tourist locations. In fact, the vegetation at this rotary gives the appearance of an impenetrable tangle. Unkempt and overgrown, it is a surprise — mostly ignored except as a functional way to direct traffic merging from four directions. It is doubtful that many of the five million or so people who drive around this circle each year on their way to the Cape Cod National Seashore have any idea of the historic legacy of the place. There is nothing to announce that they have just crossed over Cape

Cod's first canal.

The rotary was built on top of filled-in marsh. Its center is wet because the creek with two names, Jeremiah's Gutter and the more common Boat Meadow Creek, continues to flow from Cape Cod Bay, shunted via pipes under the roadway. Water flows west to east, from Cape Cod Bay to the Town Cove. Only a tide gate on the Town Cove side prevents a reverse flow from east to west.

Water and the near-sea-level elevation separating Cape Cod Bay from the Nauset estuary and Town Cove are the main features of the region, despite the effort of 1950s engineering to ignore them. During a storm, the normal ten-foot tide of Cape Cod Bay rises many feet higher, overwhelming the network of wetlands and pipes, and floods the entire area.

～≈≋≋≈～

Natural landscape and seascape change is constant on Cape Cod and is a great part of its charm. Subtle changes accrue over time, but human change often occurs swiftly and makes the landscape almost unrecognizable. A 1950s aerial photograph of the area hangs in Eastham Town Hall. The most prominent landscape feature is the extensive wetlands — a large cranberry bog, Cedar Pond, and the creeks and marshes rimming Cape Cod Bay. Just out of camera range is the site of Jeremiah's Gutter.

Today, over a half century later, all that remains of Jeremiah's Gutter is an unexceptional brackish water creek surrounded by wetlands. It is barely noticed unless there is a very high tide. The narrow shallow creek at the rotary is completely different from the prominent geologic feature it once was or the canal that Southack traveled through. Parts of Jeremiah's Gutter have filled in, disappearing into the marsh with the passage of time and lack of use, and other parts have been purposefully buried under roads, houses, railroad tracks, and other infrastructure. Only a small roadside historic marker on the auspiciously named Canal Road provides any hint of its historical or ecological significance.

Natural changes in a landscape shaped by water and geologic forces are one thing. But manmade landscape changes from development pressures with a tourist-based population have made a tremendous impact. The

**Aerial view of Orleans-Eastham rotary shortly after construction. Note pipes on right side of rotary and empty cranberry bog in center of photo. (Courtesy of Sam's Scrapbook)**

highway blocked passage for herring to get to Cedar Pond, and it also blocked the natural meanders of Jeremiah's Gutter to the Town Cove.

The cranberry bog is gone, too. Instead, a giant grocery store, a large retail clothing and housewares store, and an undulating asphalt parking lot sit on top of it. The cranberry bog was purposely filled in the 1970s prior to preventative laws. At that time, the accepted method of cranberry culture was to occasionally cover the plants with sand, and there was no distinction between sanding a bog and filling one. Many tons of sand were dumped on top of this one, but the watery peat below never created a stable or flat surface for a parking lot.

It looks like dry land, yet water is just below the surface and still flows beneath the parking lot as it always did from the old bog. It also flows from Cape Cod Bay through the remains of Jeremiah's Gutter to the rotary. Then all the water — from the bog, Jeremiah's Gutter marsh creek, and surface road drainage — flows through pipes to the Town Cove tide gate.

The maze of underground pipes directing water flow works under most conditions — but not all. A 1991 Halloween storm with its extremely high tides, immortalized by author Sebastian Junger as *The Perfect Storm*, flooded the shopping center parking lot with several feet of water. It should not have been surprising. The flooding matched predictions illustrated on flood maps issued by the Federal Emergency Management Agency. When the flood maps are superimposed over a topographic map and color-coded to show flooded land in a massive one-hundred-year storm or a catastrophic five-hundred-year storm, the amount of flood-prone land around Jeremiah's Gutter is dramatic. It covers a vast area — much of which is already developed with houses and businesses. It visually emphasizes that this is a place where water, both saltwater around it and fresh water beneath it, play important roles in land use and abuse. Moreover, those maps eerily coincide with Southack's 1720s map showing the canal, the connection between Cape Cod Bay and the Atlantic Ocean, and the Outer Cape as an island.

As sea level rises in the twenty-first century, the flood maps and the danger they detail serve as a reminder of the ephemeral nature of the Cape in general and the folly of human development in particular. Jeremiah's Gutter will turn into an inland passage to the Atlantic Ocean once again, according to a graphic projection from the Cape Cod Commission. Accessing the Outer Cape island from Eastham to Provincetown will require a bridge connection of about three quarters of a mile. Bass River will once again come close to flowing from Cape Cod Bay to Nantucket Sound.

This is the backstory to the Cape Cod of today and leads to the vision of the Cape Cod of tomorrow when water and its partner — wind — will continue to shape and reconfigure it. Ice and water made a difference in the beginning. They made a difference as people populated the land in ever increasing numbers. Water and wind continue to be relevant.

Cape Cod is an example, a case study. As sea level rises, coastal communities along the East Coast will feel the effects. Boston, much of which was built on filled land, is considering massive engineering projects to protect itself. New York, New Jersey, and Connecticut experienced what can happen because of Hurricane Sandy. Projections for Florida in the coming decades are dire.

Ultimately, water calls the shots.

.

Chapter 3
# Conflicts Seeking Resolution

The forty-foot sport fisherman *Swamp Yankee* headed out of Rock Harbor in Orleans for a beautiful sunny June Sunday on the bay. Captain Steve and his wife, Kerry, had invited me to accompany them to the 2010 annual Provincetown Blessing of the Fleet. *The Swamp Yankee* joined the flotilla of similar-sized charter boats mixed with smaller private recreational boats, all heading out of Rock Harbor as the tide ebbed. No one would be able to get back in for at least eight hours until the tide flooded again over the mile or so of flats.

We watched the boats fan out past the last of the odd channel markers — a line of trees purposely pumped into the sand flats — as we headed across the bay for the hour-plus trip to Provincetown. We looked forward to the Blessing. Always colorful, it is timed to be the capstone of the Provincetown Portuguese Festival. With the red and green flags of Portugal prominently displayed beside American flags throughout the town, the four-day festival celebrates the Portuguese culture that defined the town for generations and highlights its fishing history and tradition. The limited commercial fleet, a shadow of its former strength several decades earlier, was bedecked with flags, banners, and pennants hanging from the rigging, with all the people on board the boats in a celebratory mood.

Yet, in a town where loss of a fishing boat at sea is a harsh reality, the blessing by the bishop is solemn and taken seriously by commercial fishermen. Since 1976, fifteen men have lost their lives: seven on the *Patricia Marie* in 1976, four on the *Cap'n Bill* in 1978, three on the *Victory II* in 1984, and one on the *Twin Lights* in 2012. The Provincetown fishermen know the risks. As they pass by the pier and the bishop of Fall River sprinkles holy water on their boats, their mood becomes somber, the seriousness of the blessing acknowledging the very real perils of their profession; and they pray for safety and prosperity. After the blessing, they continue in their place in line, cheering wildly, a happy expression of gratitude and hope for the year ahead.

**Blessing of the fleet in Provincetown. (Photo by Eileen Miller)**

Provincetown has always been a fishing town. It has been and still is an art colony and safe haven for alternative lifestyles, too, but for generations fishing was its heart. Whaling, and fishing for mackerel, menhaden, cod, haddock, flounder, and sole were Provincetown's anchor.

A progression of changing boat styles and technology over time has filled the harbor with a fleet of fishing boats, from square-rigged barks and sleek schooners during the days of sail, to vessels powered by steam and gas or diesel engines including wooden Eastern-rig draggers. Later steel-hulled draggers, then smaller lobster and mixed-gear boats appeared. In the 1970s, seventy iconic Provincetown draggers, rafted four or more abreast, tied up at MacMillan wharf. By 2014, those boats had been displaced by whale-watch, charter, mixed-gear, and lobster boats, with only a handful of commercial draggers tied up there, a vivid indication of the tremendous change in the harbor.

While the Blessing of the Fleet is geared to boats docked in Provincetown, all vessels are welcome, and though the *Swamp Yankee* was not used for charters, anyone on the water knows that the unexpected and tragic happen to noncommercial watercraft, too. Captain Steve maneuvered among the boats to join a line that moved slowly in a large circle to the pier where the bishop stood. When we were parallel to the pier and directly below him, the bishop sprinkled and blessed the *Swamp Yankee* as he had the fishing vessels ahead of us.

As I looked around the expansive bay on the way home, I thought back to the Blessing. The Provincetown fishing fleet had changed so much over several decades, consistent with tumultuous changes in the fishing industry that has brought it, and the communities that support it, to their knees. The types of boats tied up at the pier and the change in the types of resources being exploited point to the vulnerability of the fishing industry. Many people who wanted nothing more than to be on the water have found substitute sources of income — hauling lobster pots, or taking passengers to watch whales, a relatively new maritime industry.

Marine mammals and endangered species represent a vulnerability of a different sort, as our exploitive practices have brought many of those populations to their knees as well. Our relationships with marine mammals, such as whales, seals, and otters — as well as endangered and threatened species like sea turtles — are intricately interwoven with the centuries-old western European attitude toward marine resources in general, which has been to take what we want. Expressing more complex sensibilities toward them is a recent phenomenon, and people are not accustomed to thinking differently about the ocean's creatures and living in concert with them. The overriding problem is that when it comes to the bounty of the sea, we want it all.

Heading back across the bay, I was struck by the number of lobster pot buoys that Steve dodged. There seemed to be an infinite variety of color combinations painted on the buoys, each marking a different fisherman's traps.

Lobstering is a prime example of the webbed interrelationships between fishing, marine mammals, and endangered species. There are two types of lobster licenses in Massachusetts: commercial and recreational. Commercial permit holders may have a maximum of eight hundred pots, and a recreational permit holder ten pots. Each license holder declares his colors on his application, brands his buoys and a plate affixed to the pot with his license number, and reports his catch annually to the Massachusetts Division of Marine Fisheries. One trap costs approximately eighty-five dollars. If a buoy is lost, anywhere from one to twenty traps can be lost with it, since pots are often set in groups, called trawls, fastened to one another.

Lobsters are the most valuable single-species marine resource and fishery caught in state waters. The Massachusetts Division of Marine Fisheries compiles commercial lobster statistics based on fishermen's catch reports for state waters, segmented into reporting districts. For Cape Cod Bay's four distinct areas, ninety thousand pots were recorded in 2014, and the bay yielded a total of 1.7 million pounds of lobsters harvested. Recreational permits are not tracked by area, but over ten thousand permits were issued. An additional 189,000 pounds of recreational harvest were reported. The direct value to fishermen was $68 million in 2014, with a far greater overall economic impact.

Lobster fishing requires bait – lots of it – and Massachusetts lobstermen use about twenty-two thousand metric tons per year of herring, menhaden, skate, and fish cuttings from processing facilities. After adding in the businesses of bait dealers, marine suppliers and lobster dealers including wholesale, retail and restaurants, trap and rope manufacturers, fuel sales, boat and engine builders, and repair people, banks and insurance companies, the industry's final 2014 impact ramped up to over $250 million. Those colorful buoys bobbing on the surface of Cape Cod Bay represent an economic powerhouse to the Cape and beyond.

Lobster fishing is the most prevalent form of fishing in the bay but certainly not the only one. Other fish and shellfish are also commercially harvested there. Charter boat fishing for bass, bluefish, and, occasionally, summer and winter flounder is also a huge business. Rock Harbor has one of the largest charter fleets in New England, a remarkable feat since it is a tidal harbor accessible only two hours before and after high tide. Fishermen dredge sea clams and hard clams called quahaugs, and they drag the bay bottom for scallops and mussels when they are prevalent. In the last couple of decades, aquaculture for oysters and quahaugs has blossomed into a new industry from Provincetown to Duxbury. The bay is constantly busy with all of these commercial and recreational fishing activities.

But fishing is just one side of the coin. The other side concerns marine mammals and endangered species like sea turtles. Cape Cod Bay is prime habitat not only for many commercially valuable species, but also for thousands of other species going up the food chain, from microscopic plankton to giant whales, that all play an equally important role in the bay's ecology and are as vulnerable to human impact, even if they aren't any fisherman's direct target.

Wrapped in strong, durable ropes so tightly that they slice through its nearly two feet of blubber to the flesh, a North Atlantic right whale is entangled and shackled in a maze of fishing gear. Although labored, it is still able to move through the water. If not freed, it may drown or die from starvation or infection months later. Speeding to the whale is a four-member disentanglement team from the Center for Coastal Studies in Provincetown. Once on site in their response vessel IBIS, team members deploy a small inflatable boat and carefully approach the whale.

They secure several lines attached to large floatation buoys to the jumble of knotted ropes, along with a control line that can be quickly released if necessary. "Kegging," as the centuries-old method is called, was once used in the heyday of commercial whaling to slow and tire out a harpooned whale so the whalers could kill it and prevent it from diving. Now the technique is adapted for a new purpose: the buoys slow the whale to save it. Then, with variously shaped sharp knives attached to very long poles, the team systematically cuts the ropes to finally get those causing the greatest damage.

It is dangerous work. The whale is disoriented, severely injured, and panicked. It does not know the team is there to help. It flails as it is slowly freed, one rope at a time. The team must anticipate what the whale is likely to do. This is a massive wild animal in distress, and one movement of its powerful flippers or fluke could be just as disastrous for the team as it was for crews of thirty-foot whaleboats over a century ago.

**Dr. Charles "Stormy" Mayo of the Center for Coastal Studies in Provincetown disentangling a whale. (Courtesy of Center for Coastal Studies)**

The team has found every manner of fishing gear entangling whales — lines with buoys attached, moorings, nets of every size and description. But locally, lobster gear predominates, likely because it is the most common. The whales get tangled up in the vertical lines leading from the pots on the bottom of the bay to the buoys on the surface, or in the lines between pots.

Virtually all of Cape Cod Bay is designated as critical habitat for North Atlantic right whales, the most endangered large whale species on the planet. It is also a feeding area for endangered and threatened tropical turtles that visit each summer. These marine animals get tangled in fishing gear, and if not cut loose, they are likely to die.

"Critical habitat" and "endangered" are terms used in two natural resource laws that motivate people to treat these animals with care. The Marine Mammal Protection Act, enacted in 1972, was designed to protect all marine mammals. A year later, the Endangered Species Act was passed to safeguard species in danger or threatened with becoming extinct. These laws, acting in concert, govern human actions with respect to the animals being protected. When there are fewer than five hundred individual North Atlantic right whales remaining in the entire world, and half of them show up in Cape Cod Bay annually to feed on microscopic copepods that flourish there, protecting them, their habitat, and access to that habitat is critical to their survival. That is how critical habitat is designated.

But conflict inevitably erupts when fishing interests bump up against such protective measures. The whales that enter Cape Cod Bay frequently get tangled in fishing gear – most often lobster pot lines. Close to one hundred thousand pots are set in that critical habitat. And when the lobster industry is worth $68 million statewide to the fishermen, including the 229 around Cape Cod Bay, and adds $250 million to the state's economy, discussions about curtailing the effects of lobster gear on whales get heated.

Compromises are adopted. Currently, lobster pots cannot be deployed until April 30, after the right whales have left the bay's feeding grounds. In 2017, that date was moved back further as the whales were still in the bay in early May, a change that affected fishermen at the beginning of their season. Yet many fishermen now work with regulators on the issues, although collaboration has not always been the case. In June 2016, the Cape Cod Times reported that the New England Aquarium received $180,000 for research to develop a rope that would be both "fishery and whale friendly," and the South Shore Lobstermen's Association received $19,000 to test a type of breakable rope. The funds were distributed from the Massachusetts Environmental Trust, made possible by the sale of right whale automobile license plates.

Entanglements lead to regulations governing fishing gear that conflict with fishermen's traditions, instinct, knowledge, and profits. The likelihood of injury or death to whales, especially the slow-moving North Atlantic right whale, also increases dramatically when whales are prevalent in shipping lanes like those off Cape Cod. Whales are no match for today's enormous commercial ships that can neither see whales nor stop or avoid them if they could see them. Altering shipping lanes and lowering speed limits reduce the problem, but those mitigation strategies can have widespread economic impacts. Preventive solutions have so far been elusive for both entanglements and ship strikes, despite years of inquiry and federal and state protections.

Between fishing gear factors, the behavior of whales, and overlapping jurisdictions, simple answers seem out of reach for these difficult issues. But these current conflicts are far from the first. The Cape's people have a long history with marine mammals and endangered species. While whales clash with fishing and shipping now, the story began centuries ago, when men from Cape Cod and elsewhere hunted whales close to home and half a world away.

# Chapter 4
## The Blubber Hunters

Earl Mills, the former Wampanoag chief named Flying Eagle, and I were both invited to participate in a 2012 book signing at Books-by-the-Sea, an Upper Cape independent book store in Osterville. He had written *Talking with the Elders of Mashpee: Memories of Earl H. Mills, Sr.*, and I was there to promote my book, *Tiggie: The Lure and Lore of Commercial Fishing in New England*.

I was grateful for the opportunity to talk to Chief Mills. When I asked him whether his ancestors hunted whales, he replied, "Sure, they hunted whales from their mishoons. It was in stories and songs."

"Mishoons?" I asked.

"Their canoes," he explained. "They used to make them from trees tall and wide enough to each hold thirty men."

"I've seen the smaller dugout canoes at Plimoth Plantation, and I saw a demonstration of making one at the Chatham Historical Society. It's a fascinating process," I said.

"Yes, it is. They could go anywhere in those canoes," he answered. "That's how they traveled." Chief Mills's comments confirmed that whales were an important resource to the native population of the Cape, and that Wampanoags possessed the nautical skills that enabled them to travel great distances in their long dugout canoes.

The process of making mishoons and their purpose was further described by Jonathan Perry of the Wampanoag Indigenous Program at Plimoth Plantation. In a video "Burning Down the Boat," produced by the plantation, Perry explained that the boats were used for trade, fishing, warfare, and general community work both locally and by ocean travel. The canoe makers found appropriately sized white pine, white oak, or chestnut trees with few branches on the lower trunks growing near the water. They peeled off the bark, burned out the base of the tree, and dropped and shaped it by fire. Then they burned the other section down to the proper size. Finally, they hollowed out the trunk. Water and clay were used to control the burn, keeping it from cracking the wood or charring it too thin. The fire worked to harden and seal the wood, preventing it from absorbing water and avoiding the problem of a waterlogged canoe.

I continued my conversation about whaling with Chief Mills. "How did they get the whales back to shore?"

"I'm not sure," he said. "Many of the details of the hunt have been lost over the centuries."

The loss of those details is unfortunate. Native American whalers are rarely given credit for the historic role they played. Traditional Wampanoag songs and stories that have survived the ages tell of the arrival of whales, generation after generation. The Wampanoags observed the patterns of whales as they migrated from birthing grounds to feeding grounds and back again in an eternal cycle. They anticipated each season and each type of the plentiful whales that would swim off their shores, and they adapted their hunting or capturing methods for the specific circumstances. Whales were a symbol of the bounty of the sea and every portion was useful, to be shared by all the villagers. As the whales appeared, the people organized hunts using implements handcrafted from items of the land and sea around them, shaped for specific purposes. After contact with Europeans, the indigenous peoples replaced rock or bone spears and projectiles with foreign tools – lances and harpoons made of metal.

The Wampanoags took advantage of whales that swam close to shore or became stranded on the flats singly or in large groups, or pods, considering them gifts sent by the Great Spirit. They drove the whales to the beach or killed them offshore. When asked how they lashed a giant whale to a canoe, Linda Coombs, former curator of the Wampanoag project at Plimoth Plantation, explained that the whale hunters worked in teams using several canoes.

Picture what it must have been like to paddle canoes large enough to hold thirty men through often treacherous waters, like those of Nauset Inlet in Orleans or Chatham Inlet, while hauling a many-ton, dead-weight whale, lashed to the crafts. It must have taken experience, skill, coordination, trust, and an immense expenditure of effort to get a massive whale to shore in paddle-powered canoes.

The Wampanoags' expertise at finding, killing, and retrieving whales close to shore was not lost on the unskilled Europeans living on the Cape. The newcomers were students of the native people, and they learned well. Whale blubber produced valuable oil, and as the demand for this precious commodity grew, early Cape colonists turned shore whaling into an important aspect of the colonial maritime economy.

From the mid-seventeenth to early eighteenth centuries, towns erected beach huts called "whale houses" on the high ground from Barnstable to Truro. Men stationed as lookouts searched the ocean for whales in the winter. At a place called Black Earth, on the beaches on the north side of what is now present-day Dennis, six-member crews from thirty-six boats, over two hundred men in all, lived in the whale houses. Hundreds more people watched and waited on the beaches lining the bay or prepared whales and their byproducts for sale. At Sandy Neck, the extensive barrier beach in Sandwich and Barnstable that separates Cape Cod Bay from Barnstable Harbor, the whale huts and lookout points were situated along the north bayside shore. The men butchered the whales, packed the chunks onto horse-drawn carts, and drove the carts over the dunes to the try yards on the south, Barnstable Harbor, side, areas set aside for rendering the blubber to obtain the oil, according to John Braiginton-Smith and Duncan Oliver in their book *Cape Cod Shore Whaling: America's First Whalemen.*

Pilot whales were a common sight in the bay. Although small by whale standards – twenty feet in length for males and sixteen feet for females – they are still bulky mammals weighing two to three tons and enveloped in a greasy blanket of blubber. The men at the whale houses scraped the blubber off the skin, commonly called "flensing," using long-handled iron tools called flensers, flat spades that were beveled on the end. They cut blubber into manageable pieces for the try pots, enormous iron cauldrons that were set over fires for many hours to release oil from the fat cells, producing liquid. It took time and heat – and required vast amounts of wood.

**Stripping blubber off a beached whale in a process called "flensing." (Courtesy of Wellfleet Historical Society)**

**Beached whales driven ashore with men working on them. (Courtesy of Wellfleet Historical Society)**

All winter long, black smoke rose along the shores of Cape Cod Bay from the fires tended by hundreds of men going about the business of processing whales, and thousands of towering trees were cut down to keep the fires burning. Blubber rendering devastated the beech, oak, and hickory forests that once anchored the soil of Cape Cod. When the trees disappeared, the soil blew away and also disappeared.

Shore whaling produced an unmistakable aroma, a primal stench wafting over the land in the stiff northwesterly winds that predominate in Cape Cod Bay after a winter storm. It was the smell of money from a lucrative business. It was so profitable that the towns claimed a share of the profits. Ownership disputes ensued if a whale was struck offshore in one town but floated ashore in another.

Demand for oil products grew in colonial times. Residents took advantage of the various whales in the waters around Cape Cod as a valuable and seemingly limitless resource. Even as the industry expanded tremendously, it was inconceivable that the numbers of whales could plummet at the hands of man.

The whale populations could not compete with man's avarice. Whale meat was disdained and discarded, but the desire for lighted cities, smokeless candles, industrial lubricants, ladies' undergarments, and a host of other commercial items proved insatiable. Industrial whaling loomed large, and Massachusetts was its center.

Long before industrial whaling started off the Cape, the Basques had been hunting whales. During the sixteenth century, Catholic dogma required nearly half the days of the year to be meatless, and although whales are mammals, they live in cold seas and were considered fish. To satisfy the demand, the courageous, enterprising, highly skilled, and secretive Basques crossed the Atlantic to a faraway land in Arctic waters, keeping their discovery a secret to avoid competition. There they hunted bowhead whales for their blubbery insulating blanket and meat, taking advantage of the tendency of bowheads to float upon death, not sink as many cetacean species do, making them easier to bring back to port.

Bowheads are baleen whales, endowed with a series of long filtering plates called whalebone, although they are not bony structures. These enormous whales, fifty to sixty feet long and weighing eighty to 110 tons, feed on tiny shrimp-like krill, filtering gigantic gulps of water through the plates and expelling the water.

Like the Basques, Cape and Islands men on board ships out on the open sea proved to be premier whalers early on. On the Cape, offshore whaling began with ships hailing mostly from Wellfleet and Provincetown. While there was no longer a market for whale meat in the European culture during the seventeenth century as there had been a century or so earlier with the Basques, the oil produced from the blubber was highly sought after. By the 1770s, Cape ports supported over a thousand whaling ships, wrote John T. Cumbler in his 2014 work *Cape Cod, An Environmental History of a Fragile Ecosystem*. Although most ships of the new industry came from the prime whaling port of Nantucket, Barnstable County records of 1690 indicate that the Islanders learned their trade from Ichabod Paddock, a Cape Codder. The Cape did not have many deep ports for the ships, but they had skilled mariners for captains and crew required by the burgeoning industry.

When they started offshore whaling, the whalers sailed close to land. As the once plentiful pods became scarcer near shore, whalers ventured farther from the coast and their home port. There, whalers encountered a sperm whale, a fortuitous discovery that changed the course of whaling and the fortunes of Nantucket and Cape Cod. The men were stunned by the enormity of the deep-diving, elusive animal, and by the quality of oil that could be extracted from it.

Sperm whales differ from bowheads in that they have teeth, lacking the baleen plates and possessing a special type of oil. Sailors often decorated the teeth of sperm whales with intricate inked carvings, an art form that became known as scrimshaw. At thirty-five to forty-five tons, sperm whales are smaller than bowheads, but they make up for the weight difference in the huge size of their head, nearly one third their overall length and containing spermaceti. This pure, semiliquid, waxy substance in the whales' head cavities produced oil for brightly-burning smokeless candles and high-quality lubrication. Spermaceti had properties better than oil rendered from blubber and was more valuable.

That treasure alone transformed the demand for whaling. Sperm whales were worth hunting wherever they could be found and whatever the cost in danger. Within a short time, investors began supplying bigger and better ships with larger crews capable of processing whale byproducts on board. These improvements enabled ships to remain at sea for years and sent the intrepid Nantucket whalers willingly out into the Atlantic Ocean in search of whales, their almost priceless oil, and profits.

They hunted whales across the North Atlantic, including the Davis Straits, where the Basques had once hunted bowheads. However, they found the frigid waters of the Labrador Sea and Baffin Bay, with their ice and bergs, far more perilous than those they had sailed before. It was a perfect home for whales. But not for whalers. The Nantucket fleet abandoned the bowheads in the Davis Straits to focus their hunt on sperm whales, and they found right whales, too.

About fifty feet long and weighing seventy tons, North Atlantic right whales are closely related to bowheads, with similar attributes that made them the "right" whale to catch despite the lack of spermaceti. They have hundreds of long baleen strips hanging from the roof of their mouths in a monstrous head similar in size to that of the sperm whale. Their blubber is a foot or more thick; they swim slowly, are buoyant, and do not sink when killed. As a preferred catch, it didn't take long for their numbers to decline.

The industry grew quickly. The whalers needed labor. Native American sailors from Canopache, the "place of peace" as the Wampanoag called Nantucket, and Noepe, "the land between the currents," the name for Martha's Vineyard as well as the Cape, had proven their extraordinary skill as expert harpooners and valued crewmembers on whaling ships. But weakened by disease and facing competition from the growing colonial settlements, the Wampanoag population on Nantucket declined precipitously after the mid-eighteenth century.

The whalers turned to Cape Cod to pick up crewmembers and expert Cape shipmasters, but it was not enough to satisfy the rapidly developing industry. As they roamed the Atlantic, the Nantucket whalers took on additional crewmembers, finding them in the Azores, Cape Verde Islands and along the African coast.

Whalers became daredevils who tempted fate. Boys who rowed the waters of the Cape for fun and adventure found all they ever wanted aboard whaling ships bound for seas far from home. A PBS documentary "Into the Deep" described the slow, sturdy, seagoing vessels as "bulldogs battling with wind, wave and whale" that lacked "grace, speed and slender feminine beauty of the clipper ship." They were state of the art and home to the sailors who signed on for "seeming endless voyages." They headed to South America and into new waters and eventually sailed east to west, through the remote Drake Passage and around the dreaded Cape Horn, a dangerous passage that remains one of the most challenging and hazardous routes in the world. They battled opposing currents and winds and monstrous seas until they could breathe a sigh of relief upon entering the Pacific. The documentary observed that the whalers were the eighteenth century's astronauts, their ships the space shuttles and rockets we know today, exploring the previously unknown realms that lay far from the eastern shores of the New World.

Whaling had become a global industry, and Nantucket became a wealthy community in the process. The island was at the center of it all – for a while. Demand grew for deep-draft vessels that could withstand the challenging conditions at sea and for home ports with harbors deep enough to accommodate the new ships. Provincetown and Nantucket fit the bill for deep harbors, with one failing. Their existence was totally dependent on the sea, a fact that did not bode well for the expansion of traditional whaling communities. Both ports had poor soil for agriculture, making it hard to feed a booming population without relying on imports. They had virtually no timber for shipbuilding and few of the much-needed, land-based enterprises needed to support the flourishing whaling industry and community. Both communities were extremely vulnerable to naval attack because they faced the open sea.

Nantucket was further plagued by sandbars protecting an inner harbor that proved too shallow for the larger ships, and it could not sustain its preeminence. The new ocean-going, deep-draft ships needed what Nantucket could not provide. Its time had passed. The industry shifted to New Bedford, west of Cape Cod, which featured a deep harbor, unlimited timber, arable land, and protection from naval attack. As the industry recovered from the War of 1812, New Bedford eclipsed Nantucket as the economic hub of whaling. So explained Eric Jay Dolin in *Leviathan* in 2006.

In the Pacific, after depleting areas close to shore and those known to be productive during previous voyages, whalers sought new areas just as they had done in the Atlantic, sailing to offshore whaling grounds further out in the ocean. No area was out of reach. Ships crisscrossed the vast Pacific, spending ever longer periods at sea to ensure full cargo holds on their return. A captain's reputation could be easily lost because of an unsuccessful trip, in which case the ship's owners would not trust him again with their investment, or the captain would find fewer experienced and able-bodied seamen ready to ship out with him.

A tour of a maritime museum in Nantucket or New Bedford; in Mystic, Connecticut; at the Makah Reservation in Washington; in Honolulu or Lahaina, Hawaii; or in Auckland, New Zealand – anywhere in the world where whaling took place – opens a door to a way of life foreign to contemporary visitors. Those museums have custody of some remaining genuine implements whalers wielded, such as harpoons, lances, the blubber-stripping flensers, and old, heavy, cast-iron try kettles, as well as photographs prominently displayed. But it is still difficult to imagine the daily existence of a whaler. Life on a nineteenth-century whaler occurred not in the black and white or sepia of photos, but in vivid color – the crimson of blood and pinkish or yellowish white of blubber, with the permeating stench and noises and heat or cold.

Visitors might look at these artifacts, perhaps thinking romantically about a life at sea, trying to envision themselves as crewmembers on board a whaler rolling on the open ocean for years at a time. They might marvel at life-size murals of whales in the Auckland museum and the skeleton or casting of one of the leviathans that hangs from the beams in

***Charles W. Morgan*, last surviving American whaler. Photo from late nineteenth century**

Lahaina. Stepping aboard the *Charles W. Morgan*, the last authentic whaleship on display at Mystic Seaport in Connecticut, it is hard to fully comprehend what it took to capture, reduce, and process the mammoths into consumer products while at sea.

Armed only with lances and hand-thrown harpoons, many whalers lost their lives in physical encounters with the behemoths as they approached sperm whales in small boats. Stories from survivors of this type of hunting are legendary, such as the "Nantucket sleigh ride" when a small, thirty-foot boat is attached by harpoon to a whale which is desperately trying to escape, taking the boat and its fishermen for a heart-pounding ride. Even reading firsthand accounts to gain some perspective about cutting the whale into manageable hunks and the scale of the operation does not make it easier to perceive.

Compellingly illuminating, and a reality check to romantic notions, is *Whale Hunt*, a book by Nelson Cole Haley, harpooner aboard the *Charles W. Morgan*. Haley shipped on for one of her last voyages in 1849 – a four-year journey circumnavigating the globe as the industry waned. Haley's account details the daily living aboard a whaler: the uncertainty and disappointment of not locating whales for weeks on end; the boredom and unending mundane chores; the euphoria of spotting a whale; the hunt from a whaleboat; the kill and the exhausting work of maneuvering the whale back to the *Morgan*; the processing; and the imminent danger.

Haley describes the flensing and trying out procedures of cutting the whale into manageable chunks and rendering the blubber for its precious oil. He wrote that once the carcass was brought to the ship's starboard side, a crewmember hung off the ship and stood on the slippery skin to secure the whale with blocks and heavy lines while danger lurked in every direction.

"Sometimes when it is rough and the whale heaving with the sea up and down, the boat steerer will be swept by the force of the sea from the whale into the sea, outside, ten or twenty feet, right among the sharks that are always around when cutting sperm whales. The man who is tending the monkey rope [a strip of canvas and rope that goes around the body as a sort of sling] at such times has to pull quick on it to get the man back before the sharks sample his legs or other parts of him..."

The work was labor intensive. It was dangerous. And it stunk. Flensing was a hard and slippery business. The long-handled tools were sharp and unwieldy. The crewmembers were elbow-deep in tons of animal fat, all the while on a pitching, rolling, square-rigged sailing vessel in the middle of the ocean.

Once whalers figured out how to keep fires going on board a vessel without burning the ship to the waterline, the onboard oil extraction became a round-the-clock operation when whales were plentiful. Rendering produced thick, black, sickening smoke, fouling the air over an otherwise pristine ocean with the same stench that wafted over Cape Cod from shore whaling. Maritime lore reveals a revulsion expressed by cargo ship sailors encountering a whaler, saying that they could see the smoke for miles but the reek upon approaching was almost unbearable. They would never approach a "blubber hunter" downwind.

Haley describes how, when working to process a whale lashed to the ship in a gale, they needed to make some cuts to let out the gas that was in the body. The whale swelled to one third more than its natural size, and if the gas was not let out, the carcass might burst open. The first cut or two with the long-handled spades sent the pent-up gases, blood, and a nauseating stench into the air. "The wind catching it and blowing it inboard gave those in line the full benefit, and soon penetrated every part of the ship."

A particularly dangerous and challenging area was the Bering Sea between Russia and Alaska, where bowhead whales were plentiful. The tremendous amount of oil in their blubber seemed worth the risk. The Arctic conditions were horrendous. Even though whalers entered the sea in the endless light of the summer, when they could hunt and process whales around the clock, the crews and ships were always threatened by unpredictable encroaching ice and devastating storms that could leave them vulnerable to the swiftly-changing seasonal conditions. A solid ocean-going ship caught in the ice became fragile. Potentially doomed, it could shift, twist, and crack apart. Rescue was not certain.

Ships that did not go to the Arctic faced other dangers. Haley's unnerving descriptions of preparing for and then sailing in a hurricane is filled with the rapid-fire orders and split-second decisions from the captain, whose "face looked worried but showed no fear," and about whom Haley said, "I hardly think from what I knew of him all the time I sailed with him, that he knew what fear was."

Haley's voyage ended when the *Morgan* arrived back in New Bedford in May 1853, carrying a mere 1,050 barrels of oil. This disappointingly low amount left most of the crew and officers in debt after four years of service. Haley had to pay two hundred dollars for his outfit and what he had drawn during the voyage, leaving him about two hundred for his four years of work. He reflected on the experience:

"The feeling of pleasure to once more be walking through the streets so familiar, cannot be expressed by me in words now. Many times I had dared not let my thoughts carry me away, as it seemed, when thinking over in the lone watches on deck at night, or swaying back and forth at the topgallant head when on the lookout for whales, that my chances of being killed by a whale or falling from aloft or tumbling overboard were as much for it as against it; and it often had appeared I never would have such pleasure again."

Haley's voyage occurred as two events led to the decline of the once-mighty whaling industry.

First was the discovery of gold. Many whaleships that sailed to California remained there, abandoned, and eventually rotting away, as crewmembers rushed to pan for gold and make their individual fortunes.

Second, oil was discovered in Titusville, Pennsylvania, in 1859. Refined petroleum quickly became the alternative to whale oil. Sperm oil prices had risen to keep pace with the costs associated with longer voyages. Abundant and cheaper petroleum illuminated the domestic needs of a growing population and lubricated the industrial revolution. In whaling's most productive year, 1847, nearly seven hundred whaling ships brought in 430,000 barrels of oil. Petroleum production delivered more than that in its first year and increased to three million barrels two years later, Dolin pointed out. The economic message was hard to ignore. The demand for whale oil suddenly disappeared. Men left their ships for steady and predictable employment on land in the burgeoning, lucrative oil fields or in hopes of striking it rich in the California gold fields

The Civil War was another blow to the industry, leaving whaling a shadow of its former glory. Ships had been burned or sunk during the war, and the fighting claimed many men. The combination of the war, plummeting whale numbers, the Arctic that took victims in well-publicized stories, and the meteoric rise of petroleum all merged to pierce the economic bubble that had been whaling. By the 1870s, the seven-hundred-ship U.S. whaling fleet had shrunk to fewer than two hundred. Whaling communities faded, and the few that survived transitioned into centers for other industries, such as fishing in New Bedford or tourism in Nantucket, claimed Eric Dolin.

Commercial whaling caused another long-term effect. It left the world's oceans with so few whales that many species were on the verge of extinction. Despite the protection afforded North Atlantic right whales through the International Convention for the Regulation of Whaling, signed by the United States in 1932 and effective in 1935, it was not until 1972 that the United States paid attention to the plight of the whale populations in general. Then they were lumped together with other marine mammals, all to be protected by the Marine Mammal Protection Act. A year later, Congress passed the Endangered Species Act.

Both laws were controversial at the time and remain so, but they changed the game for whales and other marine mammals in the United States. What was once hunted and exploited was suddenly highly protected. Now professionals risk their lives disentangling whales to save them instead of killing them.

However, even as American whaling took a back seat on the global stage, other nations were increasing their efforts. Petroleum, the substance that eliminated the need for whale oil, ironically fueled technological advances that led to whaling on a new and unprecedented level. Wind-powered square-rigged ships were replaced by diesel, steel replaced wood, hand-thrown harpoons and lances were replaced by high-powered cannons and grenade-tipped lances that exploded inside the whale's body, observed Daniel McGlynn in his 2012 article "Whale Hunting: Should Whale and Dolphin Hunting be Outlawed?" To further aid in efficiency, the largest whaling countries today, Japan, Norway, and Iceland, rely on huge factory ships traversing the world's oceans in search of whales.

The predicament of whales today is well known – and controversial. In 1946, the International Whaling Commission was established to monitor whale populations and "ensure a continued harvest." By the 1960s, though, some species were so depleted that people around the world developed concerns, and conservation efforts grew, although Japan, Norway, and Iceland continued to assert their right to hunt whales.

Acting with what they view as a matter of national sovereignty, each nation sees its own harvest as only a small portion of the total whale population. They claim that the traditional practice of whaling is linked to cultural identity and diet since a dominant portion of their economy comes from the sea. They say that their practices are "guided by scientific research and conservation principles," a claim that is often at loggerheads with conservationists. Japan maintains its whale hunting industry is related to research, and its Institute for Cetacean Research reported that 445 whales "had been killed under a scientific permit." Some who closely follow global commercial whaling disagree with that assessment and see the scientific permits as a loophole in the Convention for the Regulation of Whaling, the international agreement that established the International Whaling Commission, according to McGlynn.

In 1982, the commission imposed an international moratorium on commercial whaling. Since the moratorium, the IWC's Scientific Committee has been developing a Revised Management Procedure. While that work proceeds, Japan, Norway, and Iceland continue commercial whaling under one of two complex IWC procedures that "allow" it. Even though the scale of today's commercial whaling is a small fraction of whaling in its heyday – eight thousand whales were taken in 1853 alone – population recovery takes time, especially for such large animals, and many of the whale species are still threatened or endangered.

Conservation groups contend that the commission is ineffective because it is self-regulating, self-reporting, and lacks enforcement. Even small-scale whaling puts additional pressure on populations struggling to recover and imposes new stresses on species that can be taken with modern equipment unknown a century ago.

~~~

Aboriginal whaling is another story. Most environmental and animal welfare organizations do not campaign against aboriginal subsistence whaling, which is viewed differently in terms of scale and necessity. Some fear, however, that aboriginal whaling may open the door for groups seeking to establish community-based whaling that will eventually reestablish or increase commercial whaling. Still other groups oppose whaling for any purpose and under any circumstances on principle, maintaining that killing whales is just plain wrong.

Aboriginal whaling has "community and cultural implications," such as increasing a hunter's social status, encouraging sharing, and teaching respect for elders and their knowledge. "It's the way of life for the native Eskimos," explained Bill Hogarth, former U.S. Commissioner to the IWC. "They use [the whales] for everything, for their whole livelihood, and they share the meat between the villages," McGlynn reported.

Within this maelstrom of global debate, in 1999 the Makah tribe of northwest Washington state decided to hunt gray whales, another baleen species, from traditional canoes. Their decision followed a self-imposed seventy-year hiatus. They had stopped hunting whales in the mid-1920s when few grays swam past their region. Only when the gray whale population stabilized and the species was delisted as endangered did the Makah decide to resume their traditional hunt.

The Makah's history of whaling dates back thousands of years in a continuous record and includes the Neah Bay Treaty of 1855. That treaty granted the Makah their right of aboriginal whaling, but their exercise of that right led to emotional expressions of opinions around the globe.

Their quest was recorded in the book *A Whale Hunt* by Robert Sullivan.

He chronicled the mental, physical, and spiritual preparations necessary to succeed. The book helps explain the Makah's deep connection to, and reverence for, the whales they sought to hunt and for their centuries-old traditions. Many details of the traditional Makah hunt had been lost due to the systematic indoctrination at government Indian schools, still operating in the mid-twentieth century, so the Makah traveled to Alaska to learn specific skills from the Inuit. The story of their hunt helps to illustrate what the Cape and Islands were probably like when the Wampanoags paddled their mishoons to herd whales to shore.

However, since their successful hunt for a gray whale, the Makah have been embroiled in an unresolved court battle pitting their sovereign treaty against the Endangered Species Act, even though the gray whale population had recovered and had been delisted as endangered. The Makah continue to argue that their right to whale as granted in the established treaty will help rebuild and retain the tribe's cultural identity. Speaking for the two thousand Makah who live on Neah Bay, tribal council chairman Micah McCarty said, "We are a whaling people for some thirty-five hundred years. It's what makes us Makah."

While militant conservation groups continue to wage battles on the open seas against nations hunting whales commercially, other conservation groups have tried to persuade those whaling countries to turn to an economically viable but more benign alternative. Whale watching substitutes the camera for the harpoon. Once again, Cape Cod enters the picture as an important early model for an evolving global practice.

# Chapter 5
# Harpoons to Cameras

The first whale watch trip was scheduled for ten o'clock in the morning. A friend and I thought that would be a good time to go, before P-town, the local nickname for Provincetown, got too crazy. When we got there, the harbor was shrouded in fog. "Don't worry. It'll burn off soon," I asserted confidently, secretly hoping I was right.

As the Dolphin Fleet boat rounded Long Point, the fog began to lift, and by the time we reached Wood End lighthouse, it was gone and the sun was shining brightly. A gentle breeze rippled the sea slightly – a good omen for whale sightings.

It didn't take long. We spotted a sleek, fast-moving finback whale traveling north, parallel to us but several hundred yards farther west. It was difficult to see – a long dark convex form undulating through the water with periodic spouts shooting into the air. No dramatic tail flukes as it dove deeper – it merely disappeared into the depths below.

The real show began a few minutes later. Humpback whales were spotted a short distance ahead. Humpbacks up the ante. They seem to "play" with the boats loaded with passengers, who delight in seeing these animals just a few feet away.

Our boat moved slowly toward the whales, and the captain shifted to neutral when he got close. Three humpbacks were swimming next to the port side. All the passengers lined the rails. The whales were at the surface, lifting their massive heads every once in a while to exhale. One swam under the boat. Everyone rushed to the starboard side to see it surface. When it did, it expelled a plume of malodorous water, soaking passengers who were thrilled to be so close to the gentle giants.

The whale, his barnacle-encrusted head the equivalent of four or five passengers abreast, seemed suspended in place. His flippers were extended to keep him steady, their white undersides appearing turquoise in the cerulean water. He looked directly at us as we gazed back at him, incredulous at his immense size and calm actions. In the intensity of the moment, it was difficult not to fantasize that he was looking directly into our souls. Many of the passengers later said they felt a visceral connection to these enormous animals and could almost see a "knowing look" in the whale's eyes. They said it was an experience they would long remember.

The three whales played with the boat for probably fifteen minutes. When the whales were at a safe distance, the captain reluctantly put the vessel back in gear and inched forward, away from the animals.

**Early whale watch with head of whale close to rail of boat. (Courtesy of Center for Coastal Studies)**

**Modern whale watch boat of Dolphin fleet from Provincetown. (Courtesy of Center for Coastal Studies)**

We saw more whales further into the journey. They dove, tail flukes clearly visible, each with distinctive markings that help scientists identify individuals. This time, they were "bubble feeding," a behavior that requires the teamwork of several whales, not often observed on the boat trips. The whales create a circle of bubbles that trap their preferred food, tiny shrimp-like crustaceans called copepods and small fish like sand eels, while gulls fly and dive and squawk all around, searching for a morsel. Then the whales swim to the surface vertically, their enormous mouths open to capture the water loaded with food that their baleen plates sieve and separate. They seem to simply hang in the water as they extract their meal from the sea.

**Bubble-feeding – whales exhibit teamwork as they encircle small fish with bubbles forcing the fish to the surface where, mouths wide open, the whales have a feast. (Courtesy of Center for Coastal Studies and Dolphin Fleet Whale Watch)**

As the boat slowed again near another group of humpbacks, we were treated to a final salute. One of the whales breached sideways. The massive creature defied gravity and leapt, propelling itself nearly completely out of the water, and came back down to the sea with a tremendous splash, its flipper seeming to wave at the boat. The appreciative passengers pointed and gestured to their companions and clicked cameras to try to capture the moment.

The naturalists on board remarked that if this was our first whale watch, we didn't need to see another. We had seen the best.

Photogenic humpbacks are the darling of the whale-watching industry, but Cape Cod's waters host other species, too. One of these is a very welcome sight for those who study whales – North Atlantic right whales.

The arrival of those whales is a cause for awe and celebration. Boats full of whale watchers flood into the bay to catch a glimpse of the highly endangered leviathans at a time of year when the whale-watching boats are often sparsely populated. While vessels are required to stay at least five hundred yards away, enough aerial photos are taken for the public to be excited about having the right whales in their midst.

The fact that the right whales amass in the bay is not unusual. They come back to Cape Cod Bay nearly every year. In 2012, nearly two hundred fifty whales entered the bay, half of the estimated five hundred right whales in the world. It was proof of a modest population increase and good news for those who keep track of the endangered animals. The whales were there to chase an extraordinary bloom of zooplankton, the result of a seemingly magical synergy of conditions just right for the zooplankton population to mushroom and for the whales to exploit it, although we do not know exactly what those conditions are. It seems counter-intuitive that a seventy-ton aquatic mammal can survive and thrive on microscopically small animals that their baleen plates of specialized flexible combs sieve from the water.

Looking at a vial of the copepods preserved as an educational tool on board the vessels, it seems even more amazing that the minute shrimp-like crustaceans could possibly be the primary food of these giant animals. However, the whales, as big as they are, are improbably low on the food chain. The phytoplankton, microscopic plants, are at the bottom of the food chain as the primary producers that convert energy directly from the sun, while zooplankton, animals, are the first level of consumers. Baleen whales feed directly on the zooplankton which means they are second-level consumers with a high energy transfer. As the levels increase, the size of the animals increase while the numbers of individuals decrease. The food chain concept is often graphically represented as a pyramid with many small organisms at the bottom and few large animals at the top. A lot of energy is required for the top predators to find food in order to survive. Because there are so many animals at the lower level, it takes less energy to locate the food, so an animal like the baleen whale can efficiently convert that energy into their enormous size.

While food may be abundant, there are other threats to whale survival. As they comb the seas in search of these miniscule tidbits, they often cross paths with manmade perils with devastating results.

North Atlantic right whales crisscross a vast network of heavily trafficked maritime shipping lanes, a disastrous circumstance for any whale. Particularly dangerous places are ports in Florida and Georgia, near the right whales' calving and nursery habitats, and Stellwagen Bank off Provincetown, a prime feeding area and the shipping route to Boston. No matter how huge they may appear to tourists, many whales are small in comparison to ships. Living mostly underwater, they often go unseen and end up seriously wounded or killed by large ships.

Researchers at the Stellwagen Bank National Ocean Sanctuary traced the distribution and relative density of all baleen whales in the sanctuary and mapped the location of right whale sightings. The data showed that an area outside the shipping lanes had fewer whale sightings. As a result, and with the cooperation of the maritime transportation industry, the shipping lanes to Boston were shifted to the less-used area. Tracking the success of the compromise is nearly impossible because of the inability to constantly observe whales swimming through the area, but the sanctuary staff anticipates a significant decrease in ship strikes.

A second danger is that whales also travel slowly through fishing gear and debris – nets, lines, buoys, mooring lines – and get entangled in the gear, seriously wounding or killing them. With so few right whales in existence, every loss is a tragedy.

The news was not great in 2017. The right whales returned to Cape Cod Bay, but over a dozen of them died the following summer in Canadian waters of the Gulf of Saint Lawrence, plunging this species into an even more precarious position. The Canadian Department of Fisheries and Oceans (DFO) released a report stating that the cause of death for four whales was trauma caused by ship strikes, and one was from fishing gear entanglements. An additional four whales died in U.S. waters. In re-examining data from 1990 to 2015, Richard Pace of the National Oceanic and Atmospheric Administration's Northeast Fisheries Science Center in Woods Hole, Massachusetts, and his team found that the population peaked in 2010 at 483 whales including two hundred females. The numbers have been declining since then, and as of 2017 it was 458. Significantly, more females are dying than males. Moreover, the remaining females are having young every nine years instead of every three years as they did in the 1980s. Both factors are causing the population to decline. The Canadian Gulf was not designated as a critical right whale habitat. The sightings in 2017 were new. The report by DFO states clearly that there are knowledge gaps about habitat use, areas of concentration, and where the greatest conflicts occur between the whales and human uses. What is known is that the Gulf of Saint Lawrence plays a vital role in ship traffic through the Saint Lawrence Seaway that leads to the Great Lakes with most ships traversing the waters between Nova Scotia's Cape Breton and Newfoundland. In addition, the commercial whale exploitation maintained by a few countries also contributes to the dire situation of this critically endangered species.

The problems the right whales face are indicative of interactions

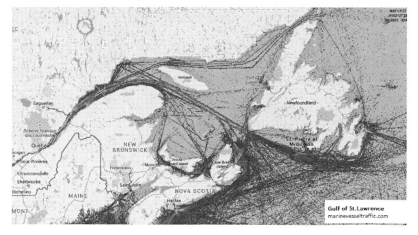

**Heavily traveled shipping lanes to the Gulf of Saint Lawrence.**

between humans and whales elsewhere, but not all of these relations are negative. Cape Cod is now a prime location for whale watching, where the inclusion of scientists aboard whale-watch tours heightens public awareness of the fragility of the whales' survival.

~~~

What turned the tide from hunting whales to cruising harbors and near-shore regions to simply watching the animals in their natural habitat? Commercial whaling had long since ceased in the United States, but in Provincetown a new and prosperous maritime whale-inspired industry took its place thanks to a serendipitous observation by an astute captain and businessman.

Al Avellar was captain of a "head boat" in Provincetown, where patrons, often tourists, were supplied with rods, reels, and bait to fish from the boat during daily excursions into the surrounding waters. Whale sightings were commonplace for him but not always for his paying guests. In the early 1970s, Al noticed that whenever his customers spotted a whale nearby, they would stop fishing to rush to the same side of the boat as the whale. One day, he advertised a trip just to observe whales and got a warm response. On the second offering, more people paid to go aboard. Before long, Avellar sold his rods and reels and began providing tours exclusively for searching for whales. He added another boat to his "Dolphin Fleet" and hired naturalists, and whale watching on Cape Cod and along the East Coast was born.

Other enterprising individuals took notice. New companies were formed, such as Portuguese Princess, and began their own whale watching tours out of Provincetown. Now boats leave from Barnstable, Plymouth, Boston, Gloucester, Salem, and other ports all over Massachusetts Bay and the Gulf of Maine, bringing thousands of patrons to watch whales every year.

Provincetown's tours remain noteworthy for two reasons. First, although observing whales from boats is the industry's principle offering, the animals frequently swim close enough to shore to be viewed from land, too, particularly at Herring Cove and Race Point Beaches. Second,

Stellwagen Bank, a National Marine Sanctuary and prime viewing area for whale watches, is within an easy boat ride of Provincetown.

The whale watching industry is now worldwide. Ports located along the migratory routes of whales around the globe cash in on the giant travelers near their shores. And for one faraway place there was an unexpected connection to Cape Cod.

*Adventure*, a whale watch vessel in Cabo San Lucas, Mexico, originally part of the Portuguese Princess fleet in Provincetown. Note the original logo depicting the shape of Cape Cod with a tail fluke at Provincetown, the tip of Cape Cod. (Photo by author)

When I approached the whale watch vessel *Adventure* in Cabo San Lucas, Mexico, in 2005, I saw an odd blue graphic emblazoned on the white cabin. It was a map of Cape Cod, with whale flukes superimposed over Provincetown. Two years later, while talking with the captain of a Portuguese Princess boat out of Provincetown, I learned that the Adventure was a former vessel in that fleet and had been sold to a company in Florida. The captain was surprised to hear that the boat was in Baja, Mexico, and that the graphic and her name endured. Provincetown, the birthplace of the New England whalewatching industry, had extended its reach to the Pacific, just as it had done with whaling over a century before.

Whale watching is now a $2.5 billion-a-year industry worldwide. In 2008, it logged thirteen million whale watchers in 119 countries, according to a report commissioned by the International Fund for Animal Welfare. The global economic potential of whale watching as an alternative to whaling indicates that the new industry could support about nineteen thousand jobs. As a center of whale watching, North America alone could gain about thirty-six hundred jobs. New England is a major attraction for whale watching. A study completed in 2000 by Hoagland and Meeks at the Marine Policy Center of Woods Hole Oceanographic Institution found Massachusetts provided 80 percent of whale-watching trips and generated $31.3 million annually. An earlier study by Hoagland and Meeks, in 1996, indicated that whale watching around Stellwagen Bank was a $440 million industry, serving over 876,000 people annually. In a tourist-based economy like Cape Cod's, the multiplier effect of whale watching is far greater. In Provincetown alone, the value of whale watching is estimated at over $2 million a year.

Whale watching provides another benefit. The Dolphin Fleet employs naturalists to provide commentary on every whale-watching voyage. While they are on board, the naturalists document the whales that they see by species, number, and, especially in the case of humpbacks, individuals identified by the distinctive markings on their tail flukes. Many of the naturalists are from the Center for Coastal Studies in Provincetown, a nonprofit organization that has been studying the local whale population since 1976 and maintains one of the longest continuous records of a whale population.

The scientists at the center have studied, photographed, identified, and catalogued hundreds of humpback whales, many that return year after year and some through several generations. It is the only institution to routinely monitor this population in both U.S. and Canadian waters. Their research combines detailed data on life history, behavior, spatial distribution, and human impacts with laboratory studies such as molecular genetics. They conduct aerial surveys when right whales are in the area, enabling them to track, identify and catalog individuals. Their disentanglement team is on call 24/7, and they train groups around the world in the methods and tools they have developed for this dangerous work. They piece together information in the snippets of time the whales are observed at the surface and use sophisticated technology to track the animals when they are beneath the surface and in their natural habitat.

Scientists from the Woods Hole Oceanographic Institution and Stellwagen Bank National Marine Sanctuary also catalog whales and even install short-term "whale-cams" to chart the courses followed by whales across Boston Harbor to the Gulf of Maine to help boats safely avoid the whales. The Division of Marine Fisheries is also involved in whale monitoring, and the New England Aquarium conducts research and sponsors whale watch activities. Thanks to these investigations, new technologies and computer applications allow fishermen, whale watch boats and other marine captains to make safe passages that may change daily, saving whales. With these highly respected institutions and the marine sanctuary conducting research, Massachusetts and especially Cape Cod are clearly at the forefront of scientific investigation into the behavior of whales, locally and much farther afield.

The change from whaling to whale watching has had a socially beneficial effect. The education passengers receive during these tours has created awareness about the marine environment in general and whales in particular that has seeped into the public consciousness, adding many voices to the conversations about the ocean. People begin to understand that low numbers of whales are of general ecological concern. The web of life is complex, but every strand is important, even if we don't understand exactly why.

In 2008, the *Charles W. Morgan*, the last U.S. whaling vessel, was hauled out of its dock at the Mystic Maritime Museum in Mystic, Connecticut, for restoration. Four years later, the ship at the center of harpooner Nelson Cole Haley's account of an 1849-53 whaling voyage in his book *Whale Hunt*, made her thirty-eighth cruise. She sailed from Mystic to Boston with stops in Newport, Vineyard Haven, New Bedford, Provincetown, and then back to New London and Mystic via the Cape Cod Canal after participating in the canal's centennial celebration. In Boston, the *Morgan* was tied next to *"Old Ironsides."* The two oldest American ships, the *Charles W. Morgan* and the *USS Constitution*, were berthed together for the first time. Although a tugboat towed her into ports of call, the *Morgan* sailed as much as possible.

**Charles W. Morgan under sail after reconstruction, July 2013.**
**(Courtesy of Dennis Murphy, Mystic Seaport Museum)**

The ship was launched on July 21, 2013, the same date of her first launch in New Bedford 172 years earlier. The historic cruise was the first time the *Morgan* had been under sail since 1921. Eighty-five "38th Voyagers," each aboard for one leg of the voyage, used their talents to "document and filter" their perspectives for the public. This group included artists, musicians, writers, scientists, actors, and historians, some of whom had personal ties to the *Morgan*. Their achievements are showcased on the Mystic website. None of her original crew members are still alive, so it was up to the twenty-first century passengers to record their own perspectives. They did this through art, music, history, and writing.

On three separate days, the *Morgan* sailed to Stellwagen Bank where the voyagers saw whales. The symbolism of what this voyage meant was deeply exhibited by the voyagers. It brought the history of whaling to the surface; to be reconciled with contemporary sensibilities. A whaleboat was launched from the *Morgan*, and the men rowed away from the ship. As the photographic record reveals, a humpback whale surfaced about fifteen feet from the boat. In another photo, a whale was accompanied by a juvenile whale, probably mother and calf. The small boat once sent from the ship to kill whales was now right beside whales, in awe and joyous celebration. The past was what it was, while the present looks toward the future. I wonder what Haley would have thought of the change in attitude and the *Morgan's* 2014 voyage.

**Whaleboat from *Charles. W. Morgan* rowing with two whales surfacing next to whaleboat, tail fluke of one and body of the other. (Courtesy of Andy Price, Mystic Seaport Museum)**

Whale watching is a business, and care must be exercised when traveling in the company of whales. But it also affords thousands of people the opportunity to marvel at whales in their natural habitat and gain some understanding of how we have interacted with these enormous creatures throughout history.

Yet, even with the economic, social, and environmental benefits of whale watching, controversy swirls around efforts to save whales and help them restore their populations. Neither commercial whaling nor aboriginal whaling has direct relevance to Cape Cod or New England today. But they exemplify the complexity of understanding when dealing with international problems such as endangered species. Often those types of problems have largescale impacts far beyond the immediate vicinity of a small geographic area like Cape Cod. Yet, when an aspect of the controversy reaches local shores, such as in the case of whales and fishing gear or whales and maritime commerce, it is worth considering the larger picture as we seek local solutions to a global problem. Entanglements with fishing gear of all types threaten the lives of whales, but restrictions placed on the use of gear create a serious problem for the fishing industry. Similarly, limits placed on maritime commerce because of whales that travel the same waters have huge economic implications as well. As restrictions are developed and then implemented, opinions often clash.

## Chapter 6
# High and Dry - Strandings

The day started like any normal early December weekday on Cape Cod in the late 1980s – cold and gray, but not horrendously windy. By midmorning, however, it became obvious that this was no ordinary day. The call came from the police. A few pilot whales had washed up on Skaket Beach in Orleans, on the inner elbow of the Cape.

A "few whales" turned out to be an understatement. The Cape Cod Bay beach seemed littered with at least fifty of them – some alive and floundering, others dead. The harbormaster rushed back to get a twenty-foot patrol boat so he could tow surviving whales that were as big as his boat into deeper water while the tide was still high. There was no time to lose. In just a couple of hours, the expansive tidal flats would be exposed, leaving the two or three-ton, jet-black animals helpless on the sand for the next eight—or more—long hours, marooned and probably doomed.

Word of the Skaket Beach strandings traveled quickly. Lots of folks were listening to police scanners to monitor what was going on in town. People arrived at the beach, some to gawk and others to help, although most had no idea what to do. They were operating on a gut response to get live whales back into the water any way they could. Without specialized training, they just reacted, trying to accomplish the goal as quickly as possible.

The harbormaster instructed his crew to tie a line around the tail fluke of a live whale, lying still in a shallow puddle, tagged to track its fate. With the other end of the rope secured to the boat, the harbormaster slowly dragged the whale to deeper water. It strained his boat, but speed was everything.

Although it seemed to be the right thing to do at the time, this approach to saving the whale turned out to be deadly. Pilot whales are designed to swim forward, headfirst. Water rushing over their blowholes from the rear drowns them. Also, dragging a sea mammal by the tail fluke stretches and probably places unbearable stress on its spinal column, much like a medieval rack would do.

A few days later, the same tagged whale washed up on another beach, dead. Of all the whales pushed or pulled back out to sea that day, pitifully few survived. Cape Cod Bay communities, saddened by the episode, were left to question why so many whales had stranded themselves on their shores and died, just as countless people over the centuries had asked the same question about mass beachings.

Rather than euphorically greeting the washed-up whales as a bountiful gift from the Great Spirit, as the Nausets perceived them, or as a financial windfall, as the colonial settlers viewed them, modern Cape residents recognized strandings as a depressingly sad and mysterious quirk of nature. Quickly following that sentiment was the realization of the financial burden involved in ridding each beach of the giant animals before they decayed, stinking up the seashore and driving away hordes of summer visitors.

Pilot whales – called "grampus" by early settlers – are a species of dolphin, not true whales. Like other dolphins, they sometimes beach themselves in groups or pods. The whole doomed cohort does not seem to sense that it is in danger. Instead, the whales mysteriously play follow-the-leader to their demise, an odd group behavior that does not enhance species survival.

While many species of marine mammals end up beached around the world, Cape Cod has the highest rate of mass strandings in this country and is recognized internationally for this phenomenon as well. The reason that these mass strandings occur consistently in the elbow of Cape Cod Bay may reside in the bay's configuration.

When the glaciers retreated, the debris they left behind continued to influence the region in four dramatic ways: the land formed a hook; the currents swirling in the semi-enclosed Cape Cod Bay formed a counterclockwise gyre; the bottom topography, or bathymetry, formed an uneven surface of shallow bars and deeper pockets; and salt marshes with wide creeks at the mouth carved a path inland toward the ocean. Individually, these four aspects are diversions; but taken together, they form a formidable barrier.

The hook-shaped peninsula traps some animals swimming within Cape Cod Bay. The hook's shank is the forty-mile stretch of glacial moraine and outwash plain from the Buzzards Bay shore to Orleans and Chatham plus the curved thirty miles to Provincetown. The barb is the buildup of sand

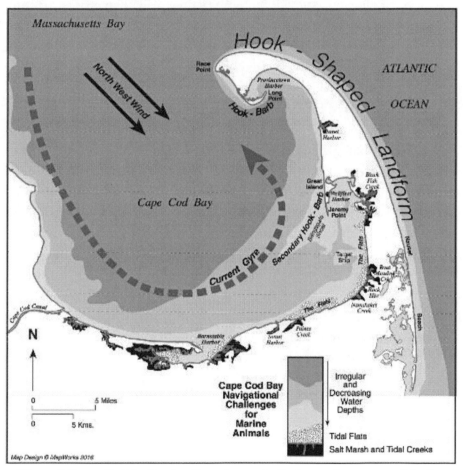

**Map of Cape Cod indicates five elements that may be responsible for strandings, singly or in combination. (Mapworks)**

deposited in Provincetown from Race Point to Long Point.

The second component of the barrier is the currents. Cape Cod forms a physical barricade, a biogeographic dividing line between the cold water currents to the north and warm currents to the south. As the cold Gulf of Maine water flows into Cape Cod Bay, it warms during the summer and circulates counterclockwise, forming a large gyre. Some of the summer animals enticed to feed within the bay may become trapped in this oceanic merry-go-round.

The third characteristic is the combination of bathymetry and depth. When glacial Lake Cape Cod Bay drained as the glaciers retreated, the sea rose. The continual erosion of the land and deposition into the bay caused by the rising sea resulted in today's bottom topography. Much of the twelve hundred-square-mile bay is between one hundred and one hundred fifty feet deep, except for the eastern third of the bay. That part of the bay features five treacherous depth changes, three shallow and two deep.

Wellfleet's Billingsgate Shoal is one of the greatest of these, acting as a secondary barb in Cape Cod's hook, much larger and more of an obstruction than the sandbar in Provincetown. When the Pilgrims landed in Provincetown Harbor in 1620, they observed the Nausets cutting up a pilot whale on the Wellfleet shoreline. At the time this area was a sixty-acre meadow, not a sandbar. Named Billingsgate for London's fish market, it was inhabited by European settlers about thirty years later, becoming home to thirty or so families of fishermen and whalers. A lighthouse was built in 1822. As the sea relentlessly beat at the island, splitting it in two, the lighthouse was rebuilt in 1855 on higher ground. Nature took its toll, and the light was destroyed by a storm a century later. Sometime in the late 1930s or early 1940s, after seawalls had failed, the island could no longer support a population. Billingsgate Island disappeared slowly into the bay.

Now, Billingsgate Shoal, a large, deceptively shallow submerged bar, extends in a triangle nearly five miles from Wellfleet in a southwesterly direction. At low tide, barely ten feet of water covers most of this sandbar. Heading south from deep water off Provincetown, a boater – or whale – comes abruptly and unexpectedly upon Billingsgate's shallow eight square miles, surrounded on two sides of the triangle by about thirty feet of water.

After crossing Billingsgate Shoal into deeper water, the next obstacle is a tongue of sand covered by only twelve feet of water that juts out from North Eastham. This smaller sunken sandbar was the final resting place for the SS *James Longstreet*, a World War II Liberty ship placed there as a target ship for pilots training at nearby Otis Air Force Base, now part of Joint Base Cape Cod. Several decades of bombing and salt exposure later, it rusted into what looked like a giant piece of Swiss cheese. For years, the hulk of rusting metal on this shallow piece of the bay served as a manmade reef, sanctuary to numerous fish species and a powerful lure for fishermen and, perhaps, for whales as well.

Shoreward of the sandbar is the second area of deeper water, a wedge-shaped channel twenty feet deep. Toward land, the last obstacle looms as a wide swath of tidal flats that line the coast from Brewster to Wellfleet and extend into the bay a mile or more from shore. At high water, the ten-foot tide covers them, and as the tide ebbs back down to zero, whales quickly lose the water depth and become stranded on the flats. These swift changes in depth, taking place a mile or more from the shore, trap animals chasing food on an ebbing tide.

The fourth bay feature affecting stranded animals is a series of marsh systems from Brewster to Wellfleet. It includes Rock Harbor that leads to Cedar Pond and was once a herring run, Boat Meadow Creek that was the entrance to Jeremiah's Gutter, and Blackfish Creek in Wellfleet that extends a mile or so inland. Blackfish Creek is named for the numerous pilot whales – blackfish – that met their fate stranded in this marsh, unable to turn back to sea. Located a few miles north of the elbow and southeast of Billingsgate Shoal, the marsh is a mile-long meandering watery indentation. From the end of the marsh creek, it is barely over another mile eastward to the open ocean. The mouth of the creek is deeper and wider than the other marsh systems, although its depth peters out as the creek heads inland. One wonders if, on an incoming ten-foot tide, the whales headed up Blackfish Creek, heeding some ancient instinctual ritual or magnetic pull toward the ocean. There they literally ran out of water and, in imminent danger, were unable to maneuver an "about face" to head back to Cape Cod Bay.

61

Pilot whales have sophisticated echolocation abilities – their own version of sonar – that make the depth issue even more puzzling. With sonar, they should be able to tell the depth of water, but something happens with their internal mechanism, which remains a mystery.

Whales are not the only animal to be tricked by Cape Cod Bay, however. The geography of the bay presents unique challenges to other marine animals.

**Marsh creeks along Cape Cod Bay. Photo shows rippled sand on the left under water but at low tide, those sand flats extend at least a mile off the beach.**

Calls started coming in to the International Fund for Animal Welfare's world headquarters in Yarmouth on a raw winter day, January 12, 2012. More than a hundred short-beaked common dolphins, smaller than the better known bottle-nosed dolphins, had beached themselves in Wellfleet.

The beached pilot whales at Skaket Beach two and a half decades earlier was a significant, eye-opening episode in the whale-saving business. The experience and lessons learned from those failed rescue attempts helped to create the Cape Cod Stranding Network, a partnership of government officials and nonprofit organizations that trains potential rescuers to handle marine mammals that continued to wash up on the shore through the decades. It is now part of the National Marine Mammal Stranding Network under NOAA, coordinated by IFAW.

This organization was put to the test yet again during the weeks of January 12 to February 16, 2012. Over the course of that month, 178 dolphins came ashore in poorly accessible and challenging marshes or muddy flats in the curve of the hook from Dennis to Wellfleet. One hundred seven of them were found dead. Others were struggling. With limited daylight, rescuers worked quickly wherever there was enough water to float the animals. Teams coaxed them off the shore and into deeper water or carried the 450-pound animals in slings to the sea.

The workforce was spread thin, responding to all the strandings. Of the seventy-one dolphins found alive, fifty-three were successfully rescued. Day after long day, teams of people trudged through the mud in the winter cold and wind to help the extraordinary number of dolphins that came ashore. The animals were marked if dead and tagged if alive. The teams recorded vital information and took samples to be analyzed in the continuous search for answers to why the animals beached themselves.

**Dolphin stranded in remote area of Wellfleet. (Courtesy of International Fund for Animal Welfare)**

Forensic necropsies – a whale-sized postmortem, and a highly unpleasant, putrid task if the decay process is advanced – are necessary investigations to document events and to determine the cause of death of marine mammals to whatever extent possible. On first examination, nothing was amiss in the dolphin incident. As the investigation continued, the mystery was never solved, commented Brian Sharpe of the International Fund for Animal Welfare.

Taken together, the hook-shaped land, gyre-type circulation, shallow and deep bathymetry, and marsh creeks could partially explain the number of strandings in the bay. Whales and dolphins might follow a food source into the shallow water on the high tide and become trapped when the tide recedes, or perhaps they search for a passage to the Atlantic that should be at this part of the bay. After all, Boat Meadow Creek – Jeremiah's Gutter – connected to the Atlantic through the Nauset estuary in the nineteenth century, and Bass River may have bisected the Cape to Nantucket Sound.

Unfortunately, the strandings continue and are not confined to mammals. Another group of annual visitors to the bay, very different from the whales and dolphins, gets caught in the hook as well.

≈≈≈

One night during the winter of 2014, Donna and Barry Tompkins received a call from the staff at Massachusetts Audubon's Wellfleet Bay Wildlife Sanctuary. Sea turtles were likely to strand. A cold front had approached, the water temperature had dropped, and the air temperature had plummeted. The Tompkins, two of seventy-five volunteers who walk the beaches, knew their night patrol was crucial, since the temperature often drops precipitously overnight and the turtles, already stressed, can succumb in such conditions. The northwest wind was fierce as the volunteers, dressed in multiple layers, headed eight or so miles from their cozy home on the south side of Harwich to their assigned section of frigid beach — Rock Harbor to Skaket Creek.

It was about midnight when they arrived near high tide. Their insulated waders added weight to their already bulky winter jackets, making walking tiresome and difficult. They walked into the wind to begin their search in order to have it at their backs on their return. While Barry waded in the ice-cold water, Donna pulled up her scarf to cover her mouth as she walked the tide line. With powerful spotlights, they painstakingly walked the high-tide line and the shallow waters, searching for the turtles that would in all likelihood be totally unresponsive, floating in the choppy water, or cast up high and dry.

Cape Cod Bay is an important summer feeding area for three endangered species of sea turtles: Kemp's ridley, leatherback, and loggerhead. Two others, hawksbill and green turtles, are rare visitors. Loggerheads and Kemp's ridley, the most highly endangered, routinely forage in Cape Cod Bay. The largest of the sea turtles, leatherbacks, weighing 650 to 1,100 pounds, are also occasional Cape visitors. All are imperiled.

Coldblooded, mostly solitary reptiles, sea turtles journey thousands of miles from tropical waters to their summer feeding grounds in the bay. Leatherbacks feed primarily on jellyfish, following their prey over thousands of miles that take them from tropical to subarctic waters and

from one side of the Atlantic to the other. The loggerheads and Kemp's ridley eat jellyfish, crabs, shellfish, and many other marine organisms but do not venture quite as far. Stomach content analyses of huge, 250-pound loggerheads have revealed sea clams and quahaugs, both of which live buried just below the surface of the sandy bottom and are common in Cape Cod Bay, especially around

**Donna Tompkins stooped over a Kemp's ridley turtle on Cape Cod Bay beach. Note clothes to indicate the time of year and weather. (Courtesy of Massachusetts Audubon Wellfleet Bay Wildlife Sanctuary)**

Billingsgate Shoal. However, as sea turtles are rarely spotted at the surface, and since they hail from warmer southern waters, few people knew until relatively recently that so many of them spent summers in the bay. Were it not for the fact that hundreds of turtles have ended up stranded on Cape shores, their presence or the bay's importance to these animals might not be recognized.

We do not know the frequency of turtle strandings on Cape beaches in the past, although Wampanoag lore tells of turtles as gifts from the sea. We do know, from several decades of documentation by the Wellfleet Bay Sanctuary, that strandings have escalated. According to Bob Prescott, the sanctuary's director, six turtles were found stranded on local beaches in 1983. A few years later, the average annual count on Cape beaches was seventeen, a number that climbed dramatically during the following decades.

In 1999, 278 sea turtles were stranded. A little more than a decade later, the 2012 tally exceeded established records when nearly 350 sea turtles were retrieved from Cape beaches. Twenty-three rare green turtles, 228 Kemp's ridleys and ninety-six loggerheads – six times the average – beached. Several loggerheads exceeded one hundred pounds apiece, and one giant weighed 125 pounds. Shocking numbers, yet no one anticipated the 2014 season.

Donna and Barry Tompkins are a link in the chain of over 175 trained volunteers who participate in the Wellfleet Bay Sanctuary's recovery program. They fight bitter cold and biting wind to walk the beaches and marshes at high tide day and night. Donna's training paid off that night in 2014. She spotted a Kemp's ridley turtle lying on the accumulated bits and pieces of vegetation in the wrack line, fine sand clinging to its flippers and hard shell. She gently picked up the barely responsive turtle and moved it to a spot above the high-tide line. There she dug a depression for the turtle and covered it with loose pieces of seaweed to insulate it from the frigid wind, much colder than the water, and marked the spot with a large rock and stick so she and Barry could retrieve it on their way back to the car.

She and Barry started up the creek, and stopped short. They had spotted a giant loggerhead, fully fifty-five pounds, in a back corner of the marsh. They covered it with grass to insulate it as best they could and continued on. They found several turtles that night to take to Wellfleet. Most beach walkers who find a turtle are instructed to mark the spot and call the sanctuary's hotline with specifics about the animal's location. The Tompkins, although volunteers for only a few years, had been thoroughly trained, and their dedication to walking the beach night and day, as well as their care in treating the turtles, earned them the privilege to transport the turtles to Wellfleet themselves, thus saving time. But the fifty-five- pound loggerhead was too big and needed extra hands to get it to the car. They called the sanctuary to report their find and give the location of the stranded giant.

Their assignment ended, they headed back to their car, picking up the turtles they had found along the way. They got to the car, cheeks red and eyes watering from the biting wind, and Donna opened a thermos of hot coffee.

Weather causes Cape turtle strandings. In spring, the coldblooded reptiles follow warm currents into the bay, but as the water cools in the fall, some turtles do not or cannot reverse the journey. The rapidly chilling water can be deadly if they cannot reach the warmth they need to

**Kemp's ridley and large loggerhead turtles rescued in Wellfleet. (Courtesy of Massachusetts Audubon Wellfleet Bay Wildlife Sanctuary)**

survive. Colder ocean water presents a physical barrier, and they linger in the bay. Once the bay water temperature drops beyond their tolerance, it is too late for the turtles. Most, if not all, of those that die are "cold stunned." Their metabolism shuts down. At that point the animals simply cannot swim, and they drift aimlessly in the bay, caught in the gyre. Fierce north and northwest winds often accompany the plunging water temperatures and drive the helpless turtles onto the beaches in the southeast corner of the bay. They land in a swath from Dennis to Wellfleet, the same area as many of the stranded whales and dolphins, making the winter wind the fifth element in Cape Cod Bay strandings.

For endangered species, the loss of every individual is critical. But rescue efforts for these animals require a lot of work by many people. In addition to the beach patrols, volunteers help at the sanctuary to pack turtles in banana boxes and transport them to the New England Aquarium's marine animal rehabilitation center in Quincy, nearly one hundred miles away. If retrieved alive, the turtles are slowly acclimated to warmer temperatures at the Quincy center. If they recover, they are released back into the wild. The U.S. Coast Guard has become a partner in the rescue efforts, transporting turtles aboard their routine flights south. Fortunately, the rehabilitation center's recovery rate continues to improve.

When the final tally was released for 2014, a shocking twelve hundred turtles were found along Cape Cod beaches and brought to the Quincy center. About two-thirds of them were alive, an astounding achievement and testament to the techniques and training of the Wellfleet team.

**A portion of the turtles found in one day. Rescued turtles were packed in banana boxes ready for transport from Wellfleet to the New England Aquarium marine rehabilitation center in Quincy, nearly one hundred miles away. Survivors, once revived, are relocated to warm southern waters and released. (Courtesy of Massachusetts Audubon Wellfleet Bay Wildlife Sanctuary)**

In January 2017, Wellfleet Bay Sanctuary sponsored a screening of the film *Saving Sea Turtles, Preventing Extinction* by filmmakers Michelle Gomes and Jennifer Ting. The film follows the life history of Kemp's ridley turtles, focusing on the plight of the turtles that strand on Cape beaches and the heroic efforts to rescue them. The two filmmakers happened to be on the Cape in 2013, became intrigued by the turtle story, and returned the following remarkable year. Narrated by renowned oceanographer Sylvia Earle, the film later aired on the Boston PBS station, WGBH.

～≋≋≋～

It was clear that 2012 was a difficult year with 178 dolphins stranded at the beginning of the year and a record-breaking 350 turtles beached at the end. The dolphin event earned Cape Cod the unfortunate and troubling distinction as one of the top three global hotspots for mass strandings of marine mammals. It shares that listing with New Zealand and Australia as places in the world with the greatest loss of life of beached aquatic mammals. Two years later, turtle strandings skyrocketed to twelve hundred. When turtles are added to the equation, the global importance of Cape Cod for animal strandings becomes even clearer.

Many questions surround these incidents, but the primary one is simply, "Why?" Geography, currents, and bathymetry are obvious factors. Cape Cod acts as a hook, trapping animals within its confines where the other elements then become important. There may be additional factors, like animal instincts or magnetic cues that are not considered or are difficult to investigate.

We can thank early cartographers who gave us a sense of what the land looked like. A century before Captain Cyprian Southack's journey, Samuel de Champlain produced a map of Nauset Harbor when he explored the Cape coastline in 1605.

Champlain's map shows several rivers flowing into the harbor. One led to the Town Cove and another headed south, an outlet that no longer exists. But an eighteenth-century map clearly shows that this southern river flowed to Nantucket Sound. When the old waterways were connected, a single inland route existed. Thus, native people and the earliest settlers could travel from Cape Cod Bay to Nantucket Sound without going around the tip of Cape Cod. By the late nineteenth century, Nantucket Sound was no longer connected to Nauset Harbor. The creek had filled in.

One wonders if boats were able to traverse this route, what about fish and marine animals? Did they utilize this waterway, too? Could it be possible that if whales went through Jeremiah's Gutter, they developed a "homing instinct" like fish that return to their birth river? If so, did they somehow imprint the former passage as a route to the Atlantic? We can only speculate. What if animals have some sort of "memory" that carries through generations? Or could there be an internal magnetic compass directing their movements, some sort of cue that signals where they should go? What if they head south, in the direction that was once the safe way to travel, and they bump into land instead? We don't know. We can speculate, but we have no answers at this point.

We watch dolphins from the safety of boats and marvel at their social interactions and their ability to ride the bow waves yet stay clear of boating dangers. We cherish stories of dolphins that guide lost mariners to safety, anthropomorphizing and even projecting mystical qualities upon these animals. We think of whales and dolphins as intelligent, sentient mammals and study them to determine the source of their remarkable skills. We

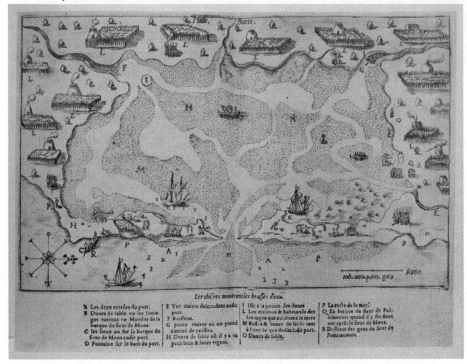

**Map of Nauset Harbor by Captain Samuel de Champlain from his voyage of 1605. The creek marked "F" may indicate a river that leads from Nauset Harbor to Pleasant Bay and eventually to Nantucket Sound. The river marked "G" probably indicates the channel to Town Cove and then to Jeremiah's Gutter. Those two rivers connect Nantucket Sound to Cape Cod Bay through an inland passage that was used by the Nausets to travel from one body of water to the other without going around the treacherous Atlantic. The harbor has changed dramatically in four hundred years.**

watch humpback whales breach and slap their fins as if waving to a crowd of onlookers, or, if fortunate, we watch them "bubble feed." We listen and call their sonic communications songs. By watching whales, we gain a tiny bit of insight into their world.

Then we see dead or dying whales, dolphins, and sea turtles helplessly stuck high and dry on a beach, huge hulks of animals that cannot exist in our element of air and out of their own – water. Modern people feel an affinity for whales and dolphins because they are, after all, mammals. We naturally want to help. Turtles, however, are not like us at all. They are not mammals but coldblooded reptiles, a throwback to ancient animals. Yet our reaction to their distress is similar: a desire to save creatures that have survived on this planet for well over one hundred million years.

It is clear that we cannot prevent strandings of marine animals. Trying to save them after they land on the beaches is probably the only assistance we can offer. Just as we cannot fully understand intricacies about the water that covers our planet and surrounds and constantly reshapes Cape Cod, so does the plight of beached animals remind us of how little we understand the creatures that live in the vast ocean. It is their home, and we are only observers.

The strandings remind us that the sea governs what happens around Cape Cod now, just as it has done in the past. The ocean provides commodities that can be exploited for profit, but it also nourishes and feeds the soul as well the body. It only depends on people's attitudes at a given time to determine which path is the more favorable for humans to follow.

Chapter 7
# Unintended Consequences

One bright, sunny, hot July Saturday, the Nauset Spit barrier beach in Orleans was filled with residents who arrived in their off-road vehicles sporting "Resident Only" stickers. I was with a group that stopped their chatter as one of them pointed at the water and exclaimed, "Oh look! It's a seal! Wow! It's checking us out! Isn't that neat?" It was a fairly unusual sight in the early 1990s.

By the end of the decade, though, seals were common along the Atlantic shore, as were grumblings from beach fishermen that landing a bass was much more difficult. The seals had chased them away. Livid longtime surfcasters could be overheard at the local tackle stores as they recounted such stories. One had hooked a sizable striped bass and was reeling it in, only to watch helplessly as a seal took the fish. Another saw a seal surface, take a bite out of the soft underbelly of a bass it held in its mouth, and leave the rest of the fish to float away. Generations of beach fishermen had been successful when bass chased sand lance toward shore into the surf zone and within casting distance. But when seals patrolled the shore outside the surf zone, the bass close to shore were trapped and seemed to learn to forage in deeper water to avoid the seals, effectively shutting down surfcasting.

Conflicts were clearly brewing.

Harbor seals and gray seals are now resident species in Cape waters. Pawns in a dynamic tug of war for well over a century, they are currently at the hub of a swirling debate on Cape Cod. One contention is that seals compete for commercially important fish. Another is that their presence brings in tourist dollars. Yet another argument considers the seals' natural predators, the great white sharks that have moved into the region, as a detriment to tourism.

Like most natural resource management issues, the navigable channel to consensus is, at best, shrouded by fog. Furthermore, the ecological role of the seals is still poorly understood. To better appreciate the complexities, we can look at a cautionary tale of unintended consequences that has parallels with the controversy surrounding the highly successful recovery of seals in Cape waters. A continent away and over a decade later, another marine mammal made a comeback.

The *Discovery*, a seventy-five-foot, seventy-five-passenger tour boat, headed out of the harbor past Land's End Inn, situated at the tip of "The Spit," an arcing low-lying glacial deposit that protrudes from the hilly terrain of Homer, Alaska. A common name for a remote point, this Land's End differs from Provincetown, also called "Land's End." Provincetown faces open ocean on its north and eastern flanks with the land mass of Massachusetts far to its west across Cape Cod Bay; Homer's Land's End faces the extreme southwestern edge of the perennially snow-covered and glaciated Chugach Mountains, home to the Kachemak Bay State Park and Wilderness. "The Spit," as it is called, is at the end of the road on the Kenai Peninsula, a dramatically different landscape from the sand of Provincetown's dunes.

On this July 2012 morning, most of the passengers on board the Discovery were Alaskan residents, familiar with the spectacular scenery yet still awed by the natural beauty. People shared binoculars and knowledge of bird species, or took photographs, or merely observed, enjoying the close-up experience, not unlike the whale-watch cruises out of Provincetown or newer seal-watch cruises on the Outer Cape.

It was too early in the season to view the orcas and humpback whales that call the bay home for part of the year, so the star that morning was the sea otter, the smallest marine mammal. Rafts, or groups, of seventy-five otters, not uncommon, brought smiles and laughter to all who watched them lazily floating on their backs, munching on treats harvested from the ocean. Mothers tended pups perched on their stomachs, feeding them crabs or shellfish they had cracked open with rocks. The constant chatter among the otters was a surprising utterance in the morning stillness. Even though the captain was careful not to approach them too closely, the otters had other ideas. Some swam right up to the boat, showing their agility, effortlessly switching from floating on their backs to adroitly diving headfirst beneath the surface to forage. Riveted and enchanted, the passengers exclaimed they were adorable.

<br>

The sea otters' story is one of resilience. Otter fur was in massive demand by the Chinese in the 1700s. While a layer of blubber insulates other marine mammals, sea otters are protected by thick fur, a double layer of six hundred thousand to a million fine hairs per square inch that traps the air that keeps the animals warm and dry in the water. These pelts were worth at least ten times the price of a beaver's, making captains and merchants wealthy. Over three hundred thousand sea otter pelts were taken between 1790 and 1818, an unsustainable exploitation that was doomed to crash, wrote Stephen Palumbi and Carolyn Sotka in *The Death and Life of Monterey Bay, A Story of Revival* in 2011).

By 1840, Chinese fur coats were no longer fashionable, the scarcity of otters made it economically unfeasible to sustain existing or build new markets and gold fever had hit. Limited hunting continued for a time. But it ceased for good because of the 1911 International Fur Seal Treaty that banned hunting for seals and sea otters, leaving about two thousand sea otters worldwide.

<br>

The loss of the sea otters had a profound impact on their habitat of kelp forests. Kelp beds are among the most productive ecosystems on earth. Fronds of the seaweed, some one hundred feet long, extend from the seafloor to the surface while shorter varieties of kelp create a vast "understory" habitat that supports a huge array of species, including shellfish like abalone and other invertebrates like sea urchins. Urchins and abalone eat kelp symbiotically: Urchins chew through the stems, while abalone eat free-floating pieces or leaves dislodged by urchins.

Sea otters are a keystone species that help maintain the diversity of kelp forests by providing a check on the urchin and abalone populations. These mammals eat fifteen to twenty pounds, one quarter of their body weight of food every day to maintain their metabolism, favoring shellfish and other invertebrates – crabs, mussels, and clams, as well as sea urchins and abalone. When sea otters are removed from this environmental equation, urchin and abalone populations skyrocket. Urchins congregate in massive numbers, strip a kelp forest bare, and then move on to other beds, leaving behind large barren areas. Meanwhile, the abalone that usually hide in rock crevices move along the seafloor to forage at will.

In Monterey Bay, California, the disappearance of the sea otters presented a case history of unintended consequences. The urchin and abalone population exploded, resulting in a cascading devastation of the kelp ecosystem. While Chinese immigrant fishermen created a profitable export fishery in the mid-1800s from the enormous numbers of abalone, the shift in species abundance caused creatures dependent on kelp for protection and food web linkages to vanish as well. When the otters were gone, the kelp beds and all the fish and animals they supported also disappeared. The ecological effect scarred Monterey Bay for nearly a century.

Dr. David Starr Jordan, an eminent fish biologist and Stanford University's first president, established the Hopkins Marine Station on Monterey Bay in 1892, patterned after the Marine Biological Laboratory in Woods Hole, Massachusetts. Dr. Julia Platt, who had also conducted research at MBL during the summers, joined Dr. Jordan in Monterey in 1899.

In Monterey, Dr. Platt became an environmental activist. She fought the famous canneries of author John Steinbeck's *Cannery Row*, vowing to reduce the pollution effects the factories had on shore, and create a space for healthy marine ecosystems. Dr. Platt established a small, ten-acre refuge for research to protect the marine life from harvest around the Hopkins laboratory and offshore. She also increased protective measures by creating a marine garden along the Pacific Grove shore where all harvesting was prohibited.

When hunting ceased in 1911, sea otters began to slowly rebound. Fifty years later, in 1962, thirty years after Julia Platt's death, they re-emerged in Monterey Bay, attracted by the abundance of urchins, abalone, and mussels in her refuge.

Without sea otters, fishing for abalone had increased, and fishermen were disgruntled when the otters returned. As sea otters foraged their way around the bay, the ravenous animals ate the abalone faster than fishermen could harvest them, threatening the fishermen's livelihood.

But the sea otters left kelp forests behind, and as the kelp became re-established, fish, seals, and seabirds returned, increasing the overall diversity of the waters. Recently, blue whales, the largest animal on the planet, have returned to Monterey Bay.

Despite these developments, Jess Righthand reported in a 2011 Smithsonian magazine article "Otters: The Picky Eaters of the Pacific" that the California sea otter population has not recovered to the extent that the northern groups in Washington, Canada and Alaska have. Terrestrial pollution, parasites and diseases washing off the land, oil in the water, and low resistance along with depleted energy of lactating mothers in areas of limited food all contribute to the poor recovery. Others have found an increasing problem with sharks. Sea otters occasionally show up dead on the shore with puncture wounds from encounters with great whites. The predators seem to bite but not eat the otters, and while the mammals are not killed immediately, the attacks still lead to their death. No one knows why sharks do not devour the otters outright. Maybe they prefer high caloric blubber over muscle, but perhaps those six hundred thousand hairs per square inch are not palatable. Now that sea otters are protected, their exquisite fur coats keep them warm, not us.

Great white sharks are not prevalent in Alaska's Kachemak Bay, but even sea otters there are inexplicably dying faster than normal. Sea otters are now seen as an indicator species as well as a keystone one, given the myriad environmental connections surrounding them. If their population is healthy, the ocean probably is too; but research suggests that the reverse seems to be the case now.

The story of Monterey Bay is instructive as we grapple with expanding populations of seals on the East Coast. Seals were a rarity around the Cape and Islands prior to the late 1980s. A century earlier, seals were prevalent, hunted and harassed to nearly total extirpation. Now, there are thousands. The population of seals has skyrocketed in Chatham and Nantucket, which are the focus of attention as significant seal pupping and feeding stations. While seal-watch companies have sprung up as a relatively new tourist attraction, not everyone is as thrilled as the tourists to see them return to Cape waters.

Basques occasionally took seals off Newfoundland during the sixteenth century, along with whales and cod, but sealing began in earnest in the 1720s and expanded over the decades with the fur trade. In addition to otter pelts, less profitable seal skins sold in Canton for a dollar or two. Volume compensated for price. Seals congregate to rest on exposed sand bars or rock outcroppings or ice depending on the circumstances in what are known as haul-out areas, and the ease of clubbing them to death was astounding. A century later, half a million seals were harvested annually, and by the late nineteenth century, sealing was second only to cod fishing as Newfoundland's most important industry. With the boom in seal skin trade – reportedly one hundred thousand in a single voyage – their numbers dwindled along with the otters'.

"Culling" harp and hooded seals, which require ice-jammed waters for their survival, continues in Canada where the seal harvest still plays an economic role as an important export. The practice is certainly not without contention, though, as sensibilities about marine mammals and their role in the environment have evolved.

In 1972, when the U.S. Marine Mammal Protection Act went into effect, sea otters, seals, whales, and dolphins could no longer be purposely harmed. Unlike the Endangered Species Act, there is no provision for species recovery plans or population manipulations.

Harbor seals spend their summers in the Gulf of Maine and swim south to Massachusetts and New York's Long Island in the winter. Seals come and go from dry land to water. Surveys made at haul-out sites where the seal presence is fluid are extrapolated to produce a scientifically acceptable estimate of the population, states Lisa Sette, seal research coordinator at the Center for Coastal Studies A count of thirty-eight thousand harbor seals in Maine in 2001 was estimated to represent a larger population of about one hundred thousand animals.

**Seals hauled out on sandbar in Chatham Harbor. (Courtesy of Blue Claw Boat Tours)**

Though gray seals were slower to repopulate Massachusetts waters, researchers believe their numbers also increased substantially: from a few in 1974 and 1975, to between five hundred and a thousand in 1993. By 2001, the number had grown to between fifteen hundred and seventeen hundred. The Northeast Fisheries Center estimated in 2009 that one thousand gray seal pups are born annually on Muskeget Island on the southwestern tip of Nantucket. Over three thousand seals were estimated to congregate on the island in the winter, indicating a rapid population escalation and year-round residence. In 2012, the Massachusetts Division of Marine Fisheries revised those figures impressively upward: two thousand pups per year on Muskeget Island and fifteen thousand gray seals in Cape and Island waters. In 2017, utilizing satellite imagery, Dr. David Johnston estimated the total for gray seals at between thirty thousand and fifty thousand, living mostly around the Cape and Nantucket waters. Yet that is considered to be a small segment of the total population.

At the twentieth anniversary of the Cape Cod Natural History Conference in 2015, Stephanie Wood, from the National Oceanic Atmospheric Administration's Northeast Fisheries Science Center, presented a retrospective on the seals. She explained that on Sable Island in Nova Scotia, researchers looked at biological factors such as age at first production being about three years old, and the juvenile survival rate, both of which indicate a strong and increasing or stable population. When either the age for first production increases or the juvenile survival rate decreases, it indicates the carrying capacity has been exceeded, and the population begins to decline. The researchers don't yet know what the local carrying capacity is.

Sable Island supports fifteen thousand gray seal pups per year, nearly eight-fold more than on Nantucket, and hundreds of thousands of adult seals have been spotted in the Canadian Maritimes. While the numbers of pups and adults are increasing around the Cape, those are just a fraction of this entire population that extends from Labrador to Long Island. Harbor seals also have a wide geographic distribution – from Davis Straits to New Jersey. Tracking individuals shows that they swim throughout their range, so any seal count is only a representative sample of the population. It is not a "resident population" that stays in one area. Individual seals present on the Cape on any particular day may be in Canada's Maritimes a few days later.

Such a large geographic distribution for both species crosses state, provincial, and national boundaries, and their habitat overlap means that one or the other species is around Cape Cod year round. The mobility factor of such a large group of animals means that "managing" them is exceedingly difficult, and the consequences of their presence are complex.

These seals eat herring, cod, and flounder, all commercially important species, along with many other species, including striped bass and sand lance. Each adult seal consumes twenty-two hundred to twenty-six hundred pounds per year. With an estimated one hundred thousand harbor seals in Maine, there is the potential of seals eating a hundred thousand tons of fish per year.

The local Cape populace did not recognize the amount of food that could be consumed when the Marine Mammal Protection Act went into effect because most seals were north of Massachusetts waters. But when harbor seals began to show up on Cape Cod and the Islands in the late 1980s, first around Nantucket and then the Outer Cape, awareness of their fish consumption increased. The more recent gray seal population, the huge amount of fish the seals consume, and the impact on Cape area fisheries have attracted the attention of fishery managers at all levels as gray seals have become year-round residents.

It is not a new dilemma. The relationship between fishermen, seals and fish, and predators and prey has been at the heart of one aspect of the debate over seals that years ago led to a bounty system.

**Gray seals in Chatham Harbor. (Courtesy of Blue Claw Boat Tours)**

For nearly one hundred years, both Maine and Massachusetts had a controversial management scheme, a bounty system, to "cull" seals that competed with fishermen for fish. The themes of fishing, tourism, and economics in those arguments over a century ago have been echoed in current comments heard on the street and reported in local newspapers today.

When trains and steamships improved transportation shortly after the

Civil War, Maine became a fashionable vacation destination. Opulent homes, quaintly called "cottages," and grand waterfront hotels lined the picturesque coast for well-heeled out-of-staters. Killing seals to protect fishing interests was practically a daily occurrence – to the disdain of the wealthy waterfront residents and vacationers, who expressed how appalled they were in letters and articles in the local press. They wanted laws that prohibited hunting, stated Barbara Lelli and David Harris in their 2006 paper "Seal Bounty and Seal Protection Laws in Maine 1872-1972: Historic Perspectives on a Current Controversy."

Animosity resulted. Seal hunting was banned in Penobscot Bay in 1872 and seals were protected for the two years that the ban lasted. Three years later, in 1877, the shore and the inhabited islands in Casco Bay, home to Maine's largest beaches, gained protection. Residents argued that "seals were attractive" to them and to tourists. Fishermen in Harpswell argued for an exemption from the prohibition of killing seals, clearly drawing the lines between fishermen and shore-side residents and visitors.

In 1891, the fishing industry influenced Maine legislators to enact a seal bounty for Penobscot Bay that also effectively repealed the Casco Bay ban. To receive the bounties of fifty cents per seal, hunters presented seals' noses to town treasurers, who paid the bounties. The treasurer disposed of the noses and the state reimbursed the town.

A see-saw relationship developed along the coast between the groups with divergent viewpoints. Hunting bans drew complaints from fishermen, and other residents expressed the opposite reaction when hunting was approved. In 1905, the payout was increased to a dollar as incentive to kill seals, but the bounty program was soon repealed again, condemned as useless and expensive.

For nearly thirty years, there were unsuccessful attempts to revive the sanctioned killing. Then in 1929 an act was proposed that generated considerable controversy. It instructed the "gunner" to provide the town treasurer with the whole head of the seal, not just the nose. The language for and against this proposal was colorful, as one might imagine. Those in favor spoke of the conflict between seals and fish resources. But picture a town treasurer presented with a seal's head that must be disposed of, while the headless carcass floated by a fancy residence or grand hotel or was left on the shore to rot. The measure ultimately failed.

The bounty issue continued back and forth between resident and fishery interests, and by 1945, communities with large tourist industries were pitted against communities with major seal and fish interactions. Environmental groups began to weigh in on the anti-bounty side of the issue, and the seal head proposal was revived as well. The bounty law was repealed, and responsibility for the control of seals was transferred to the commissioner of Sea and Shore Fisheries. He was directed to "kill and dispose of all seals in the waters of any of the coastal counties of the state whenever such seals are causing damage to property or livelihood of fishermen."

In 1947, the department interviewed fishermen in eighteen communities to determine the extent of damage to nets and gear and losses in the amount of fish eaten by the seals, schools that had been broken up, and fish driven away from their gear over a five-year period. The vast majority of responses revealed that fishermen felt seal controls were necessary and favored a bounty, but 36 percent reported no damage to their gear, and 61 percent reported negligible or no losses from seals eating fish. Fishermen wanted a bounty, but they had not provided evidence to the commissioner that seals caused damage to commercial fisheries. No clear picture emerged of exactly what was taking place. A separate study revealed seals ate herring, flounder, squid, skate, and sculpin but it was inconclusive regarding actual losses to the fisheries as a whole.

The last call for a seal bounty was defeated in 1962. A decade later, seals came under the protection of the federal Marine Mammal Protection Act that preempted any state laws. Prior to enactment of the MMPA, the U.S. Congress Merchant Marine and Fisheries Committee issued a report stating:

"Recent history indicates that man's impact upon marine mammals has ranged from what might be termed benign neglect to virtual genocide. These animals, including whales, porpoises, seals, sea otters, polar bears, manatees, and others, have only rarely benefited from our interest: they have been shot, blown up, clubbed to death, run down by boats, poisoned, and exposed to a multitude of other indignities, all in the interests of profit or recreation, with little or no consideration of the potential impact of these activities on the animal populations involved." (U.S. Congress Merchant Marine and Fisheries Committee Report 1971b: 11-12).

The Massachusetts story is similar to Maine's though not as well documented. A seal bounty was enacted in 1888. In 2009, Barbara Lelli and David E. Harris examined the century of historic and social issues surrounding seal bounties and protection laws in both states in "Seal Bounties in Maine and Massachusetts, 1888-1962." They estimated that between 72,000 and 136,000 seals were killed merely for the bounty and that most were killed in areas where both seals and humans interacted in greater concentrations.

Seals, fish, fishermen, tourists, the economy, aesthetics, safety, and a type of morality all played a part in the bounty controversy, yet there is no clear determination of the effect that the bounty had on the seal population as a whole or on the fishing industry. Little has changed after a century and a half. The debate, arguments for and against control of seals and parties involved, continues.

In 2009, the Woods Hole Oceanographic Institution sponsored a Gulf of Maine seal conference. Those involved recognized that the bounties, while an early attempt at ecosystem management to reduce predatory pressure and restore depleted fish stocks, also competed with seal protection influenced by a burgeoning tourist industry. Those basic attitudes and values have remained at the root of the conflicts over the past century.

The Marine Mammal Protection Act seeks to conserve and protect all marine mammals. It says: "Populations ... should be protected and encouraged to develop to the greatest extent feasible commensurate with sound policies of resource management and that the primary objective of their management should be to maintain the health and stability of the marine ecosystem." This reasonable-sounding goal forms a basis for restoring and protecting depleted populations. However, it omits any definition of the "greatest extent feasible for population development," or "precise and measurable indicators of ecosystem health," or how to address stable or increasing populations. These are not trivial concerns.

Ecologically, the lower limits of a population's viability can be measured by indicators such as recruitment, the rate at which juveniles join their parents. The upper extent of "feasible" population growth is more difficult to define and is usually more subjective. How can "too many" be determined?

Three main questions, and the sentiments these questions engender, fuel the seal debate: How is "the greatest extent feasible" defined for seal populations? What are the precise and measurable indicators of ecosystem health? How many is too many?

Data is needed to document the seal-fishery interaction, but a method to obtain the information remains elusive. Two years after the Woods Hole conference, the Center for Coastal Studies sponsored a follow-up meeting in 2011 and invited fishermen to attend. I talked to a few local beach fishermen at the time, asking if they were aware of the meeting and if they were going. They bluntly expressed the sentiment that attending would be a waste of time because in the end "nothing is ever done" to address the issues.

The fishermen who did attend vividly described seal interactions. Hook fishermen recounted incidents of seal herds that followed their boats and stripped the hooks of any fish that had been caught before the fishermen could get them on board. Net fishermen told of shredded gear. Weir fishermen explained that while fish cannot find their way out of a weir, clever seals had no trouble getting in and decimated the catch trapped inside. The protestations may not have included the measurable statistics favored by researchers, but absent of such data, the fishermen's anecdotal observations were compelling and exemplified the complexity of human-seal interactions.

A 2011 report by the National Oceanic and Atmospheric Administration states the obvious: "Long-standing conflicts between humans and marine mammals, particularly seals, relate to marine mammal ecological impacts on economically valuable fishery resources. Most

84

marine mammals are apex predators, removing tons of prey from the ecosystem." Less obvious, the report continues, "Some estimates equate the level of removal to equal that or exceeded by fisheries." In other words, seals may eat more fish than fishermen harvest.

The 1972 Marine Mammal Protection Act did not provide for a management or recovery plan or determine what balanced interests might be. Instead, it stipulated protecting all marine mammals from killing, harassment, or harm, and instituted stiff fines for anyone caught illegally molesting any marine mammal. Four years later, in 1976, the Magnuson-Stevens Act was passed. It enacted a coastal-wide, two-hundred-mile exclusive economic zone and established the fisheries management schemes that have been in existence – and vociferously fought about – ever since.

Adding to the complexity of the seal controversy, NOAA is charged with enforcing both the Marine Mammal Protection Act and managing commercial fisheries. Fish are currently managed species by species through specific management plans to promote recovery of depleted stocks. Those plans do not necessarily consider the role of a species in relationship to other species, or habitat overlap, or other common factors. Conversely, marine mammals are managed for complete protection. With these different and sometimes conflicting management schemes, natural food web interdependencies are often overlooked, and maintaining consistent policies for fisheries and mammals is challenging.

The issue of cod demonstrates an important link between the two laws. The cod population was very low when the fisheries act was enacted, and building stocks was and still is a primary objective. In the course of that effort, fishermen have been forced to change the way they fish, how much fish they can catch, where they can catch them, and how many days during a year they can fish. They must also file seemingly endless paperwork to document their activities. At the same time, marine mammals, also low in abundance when the MMPA was passed, were protected, and their populations began to expand.

Seals eat cod. They also eat many other fish species. NOAA recognizes the irony of the dichotomous charges to manage both types of resources so differently, and it recognizes the benefit of ecosystem-based management. During the early years of both of these laws, seals were not considered a threat to the fish. But as the decades have passed, cod stocks have not rebounded as hoped, while the seal populations in the Gulf of Maine have mushroomed. These two separate occurrences collide headlong with one another as fishermen continue to watch seals catch fish that they want to take to the pier themselves. It all renews pleas that "we have to do something about seals," an echo of a century ago. When climate change is added to the mix, everything becomes more complicated.

For many people, there is another compelling reason to "do something" now – the terror of great white sharks.

# Chapter 8
# Sharks Take a Bite

When Orleans residents arrived at Nauset Spit to share holiday festivities on a hot and sunny 2012 Labor Day, the water temperature was in the high sixties, unusually warm for the Atlantic Ocean. It was a perfect day to bask in the sunshine and cool off in the surf. Yet no one along that stretch of beach dared venture into the water beyond knee depth. The public beach half a mile south had been closed to bathers earlier in the week. Large great white sharks had been sighted close to the beach.

People at the "swim at your own risk" beach at the spit heeded that warning. A few weeks earlier, a vacationer from Colorado had been attacked by a great white shark fifteen miles north in Truro. All day, people pointed to seals poking their heads above water to watch the people watching them. Then those on the beach searched for telltale fins. Gray seals are a favorite food of great white sharks, the reason the sharks were there.

As seal populations have increased, so has the presence of sharks along the Cape Cod shoreline. The interaction between the two predators, the role of gray seals switched from predatory fish-eaters to prey, is an issue of increasing concern among resource managers.

Representatives from Outer Cape towns and the Cape Cod National Seashore gathered to discuss the occurrence of great white sharks and what their response should be. They acknowledged that everything was different, and they knew the beach would never be the same.

Dr. Greg Skomal, shark expert for the Massachusetts Division of Marine Fisheries, noted there hadn't been a great white shark attack in

Massachusetts in seventy-six years until that summer.

As a tourist destination, officials were not only concerned about public safety and potential serious injury. They were also concerned about the impact on tourism.

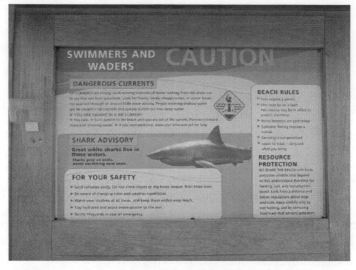

**Beware of possible sharks in the water kiosk sign.**
**(Photo by author)**

As the seal-shark occurrences gained attention in the media, reports included statements about the limited access to the pleasures of the beach: signs indicating dangerous marine life in the water, people permitted to only wet their feet, and lifeguards asking waders in deeper water to come ashore. The media coverage of the great white sharks sent a chill through Orleans during that perfect Labor Day weekend in 2012. At that traditional end of summer, usually a hugely popular beach weekend, only a hundred cars were parked in the nine-hundred-space lot because people could not play in the water.

In the give-and-take between humans and sharks and other marine life, Cape towns face a new reality. The beaches are vital to the tourist-based economy, and beach closures, whatever the reason, cause an economic ripple through the community. Outer Cape beaches are among the most popular in the country. The Cape Cod National Seashore is one of the top ten national parks, with five million visitors a year; and Eastham's Coast

Guard Beach was ranked among the top ten beaches nationwide from 2010 through 2018 by Dr. Steven Leatherman, known as "Dr. Beach." Nauset Beach is world renowned, and Chatham's Lighthouse Beach has grown in popularity since it was created in 1987 when a storm broke through Nauset Beach and formed a new inlet.

Orleans and the National Seashore charge fees, while Chatham's beach is free. For Orleans, Nauset Beach is, or has been, an economic boon to the town's coffers. But with great white sharks in the area, officials wonder if people will still go to their beaches or find another seashore or discover a different summer diversion. It is a sobering economic concern for the towns.

Towns and the Cape Cod National Seashore are faced with many issues, most related to public safety and funding. With the rise in shark sightings have come new concerns: Whether there would be funds for rescue equipment such as jet skis rather than traditional surfboards, upgrading training for first responders, emergency medical supplies at the beaches, a coordinated communication system across jurisdictions for shark sightings, and legal funds for potential lawsuits and assistance to help avoid legal entanglements. Also, who or what should determine when it is safe to swim? Selectmen from the affected towns requested an investigation into how other parts of the world, such as South Africa and Australia, deal with sharks.

This problem is not as simple as finding a way to "adjust," or drastically reduce, the number of seals to get rid of the sharks. Great white sharks, apex predators at the top of the food web, also serve an important ecological role of helping to keep other marine populations in check.

Those sharks are in Cape waters because of the gray seals. If we try to decrease the number of seals because of the sharks they draw to the area or because of the fish they eat, we do not know what the consequences would be. Would other species be affected, or would changes in the population in one area impact other environments where the seals are prominent? Would there be cascading effects in the food chain? The story of the sea otters reminds us that manipulating marine populations to suit our purposes through one policy or another – with unknown and unintended environmental, social, or economic consequences – is what we have always done.

The forty years since the Marine Mammal Protection Act was passed may seem a long time to humans, given our well-known demand for instant gratification, but it is a short time in terms of large mammal population dynamics. Seals have made a strong comeback, but eventually they will exceed the environment's capacity to support them if natural forces prevail. Disease, predator-prey interactions, food requirements, or some other occurrence will check their population growth. It would stabilize if a hands-off policy were implemented which is something we never seem to do.

**Great white shark**

The economics surrounding seals extend in many directions. Seals eat fish that fishermen anticipate marketing. Seal-watch tour boats now bring thousands of tourists to haul-out areas and surrounding environs. The increasing gray seal population also means we can expect more great white sharks around, since the seals' only predators are sharks and man. The sharks are redefining the roles of town beach managers and first responders to protect the public at a substantial public cost and a simultaneous loss of anticipated revenue from the lack of people paying to use the beaches. Services that cater to tourists have decreased business when fewer people pass by on their way to the beach. All of these interests weigh in on any discussions of seals.

Late in the summer of 2012, a new economic enterprise appeared that may have caused longtime Cape Codders to chuckle. A startup industry began to advertise "shark tours." One hopes they won't "need a bigger boat," to borrow a line from *Jaws*.

One early fall day in 2012, as I rowed around Pleasant Bay, I was startled by a shiny, smooth black head popping out of the water, not more than fifteen feet in my wake. The only sound was a muffled swish as the seal exhaled. Curious, he watched me stroke rhythmically, slipping below the water and reappearing every once in a while until I ventured into shallow water. While thirty years ago his presence would have been a rarity, this close encounter with a single seal was not uncommon. I looked at him thinking that he and his comrades were creating quite a stir.

Seal-spotting and whale-watching tours. Lobstermen, cod fishermen, and bass anglers. Ships and maritime commerce. Beach-goers and the industry that supports their presence. As men and women harvest commercially or recreationally from the sea, or transport goods across the oceans, or merely go to the seashore to swim on a hot summer day, we are confronted with a new reality in how we use the seas.

**Seal tour boat (Courtesy of Blue Claw Boat Tours)**

In the past, men traveled the world's seas hunting oceanic animals, or local hunters were paid for killing them. Now, marine mammals and endangered species such as sea turtles have been protected since the 1970s, when the U.S. Marine Mammal Protection Act and Endangered Species Act went into effect.

As a result of these laws, critical habitat has been defined and regulations enacted to protect those areas; fishing gear has been modified to prevent entanglements of whales and turtles; and shipping lanes have been modified. Hunting bans led to a huge increase in Canadian Maritime and New England seal populations, and the consequent economic fallout from the seals' impact on fish harvests, and the appearance of great white sharks that prey on them in turn, spreads in many directions.

Our collective attitudes toward marine mammals are fickle. At the same time Cape fishermen were decrying the increase in seals and decrease in their fish catches, another drama unfolding only a few miles away elicited a completely different response. When 178 dolphins stranded on the Dennis to Wellfleet beaches in 2012, people on the Cape expressed sadness at the senselessness.

Humans have a positive emotional attachment to dolphins, no doubt in part because of popular films and television programs portraying these animals as intelligent, gentle, and almost humanlike in their behavior. This warm, nuanced reaction is far from the straightforward negativity of those whose personal bottom line and livelihood are affected when seals do what they do naturally – eat fish.

Dolphins also eat fish, but there is no widespread animosity toward them, and the amount of fish that dolphins eat does not seem to be factored into fishery statistics. Instead, in a compassionate response, people reel when seeing dolphins or whales in distress. If whales or turtles are entangled in a maze of fishing gear, then we try to cut them loose to protect them. If animals like pilot whales, dolphins, or turtles strand on a beach, high and dry, then the focus is to try to get them back to their natural habitat, to save them from death – and to save the community the cost of disposing of their enormous bodies. And in a bizarre twist, people feel the need to help even a stranded great white shark and work feverishly to get the large predator in the water, as has happened twice since 2014.

Our attitude toward marine species depends on how they interact with us and whether they affect us directly, primarily through economics. Not surprisingly, the people who depend on fishing for their livelihood are generally not the same people who comb beaches searching for stranded animals. Non-fishing businesses create livelihoods, too, by ferrying people out to where the animals live. Whale watch, seal tours and now shark tours give people an opportunity to see animals up close in their natural habitat, an experience often remembered long afterwards. These businesses do not compete with the animals for what they do or don't do in their natural surroundings. Just observing them provides the economic incentive.

However, if regulations regarding marine animals diminish peoples' ability to make a living or adds substantial costs and reduces the benefits to that livelihood, then stakeholders will implore regulators to "do something" to lessen that loss. If "doing something" means limiting the animal population, then defining how many is too many is essential. But the current law does not allow for that type of management flexibility.

Fishermen know the general public has an emotional attachment to marine mammals, including seals. At an open forum in Chatham in 2011, John Bullard, then the newly-appointed director of NOAA'S Northeast Fisheries Center, and former mayor of New Bedford, Massachusetts, was on the spot as he answered policy questions. As one fisherman asked, "Who would be the one to suggest going back to a bounty?" Several wondered what was going to be done or could be done about the seals. However, like other such officials, Bullard was restricted by the limitations of the Marine Mammal Protection Act, the law he was sworn to uphold. The implication was clear. Unless the law was amended, there was nothing that Bullard or the agency could do except perhaps try to bring parties together to determine the definitions of populations and healthy marine ecosystems.

How to accomplish that goal remains obscure. As the controversy continues, we still want it all. We want fish and lobster and other seafood at home and in our restaurants, and we want fishing communities to supply that seafood. We want to surf-cast off the beach or troll for fish from our recreational boats. We want to watch giant whales leap out of the water and come back down with a tremendous splash. We want to see seals popping their heads out of the water, seemingly checking us out, but we don't want them eating the fish that we want for ourselves. We don't want to see turtles or whales or dolphins stranded on our beaches, and we want to find ways to stop this ancient phenomenon. We don't want to see animals suffer and struggle to survive after being literally tied in knots. And we want to swim safely in the water without worrying about seeing a shark fin nearby.

So we are left wondering. How can we have it all including a vibrant lobster fishery without further endangering right whales; or a recovered ground-fish fishery in the face of an exploding seal population? Will we learn to live with both seals and sharks threatening the public? Or will we once again manipulate marine mammal populations to suit our own purposes and pocketbooks; and watch the drama of unintended consequences unfold? How do we maintain both a desire for economic prosperity from marine resources and a balanced ecosystem?

There is no crystal ball for this debate, and at this point those issues are being deliberated separately. But the concept now gaining broad-based acceptance is that, ultimately, we need to find ways to investigate and integrate all the parts of the puzzle on an ecosystem-wide basis. That concept is being explored by NOAA in a draft report for managing resources on Georges Bank. International experts are expected to present an assessment on the draft report in 2018.

Chapter 9
# The "Tragedy of the Commons"

The gleaming stainless steel machines commanded attention. Conversation was impossible anywhere near them, drowned out by deafening metal-on-metal grinding and clanging. Pockets of people crowded around each machine at the 1976 International Boston Seafood Show in the cavernous Hynes Memorial Convention Center.

At one, the operator, outfitted with a black plastic apron, placed a two-foot headed and gutted cod on a conveyor belt to be filleted by the White Fish Filleting Machine, as it was called. Two perfectly-cut, boneless fillets emerged mere seconds later. Another machine filleted flounder, while a skinning machine made short work of cutting the skin from both the cod and flounder. Hanging from the steel girders above, a banner advertised the company's name in large bold letters: Baader®. The German company, the first to develop a heading and deboning machine for herring in 1921, had grown as an industry leader to also produce the first salmon and cod lines of machines. Occupying highly visible, centrally located, and valuable exhibition floor space, Baader® had invested heavily in the show.

A short distance away, salesmen demonstrated how quickly their enormous freezers used flash-freezing technology and showed potential customers the end product of frozen fish fillets. Some products were vacuum-packed and some came out ready to be packaged.

As a first-time visitor to the annual seafood show, as the new shellfish biologist for the town of Orleans, I was overwhelmed by the astonishing variety of machines, the array of seafood vendors, and the representations of every possible aspect of the seafood industry. Sampling delicious seafood from companies involved with consumer products was a tasty bonus.

I did not know it at the time, but 1976 was an important year to begin observing the fishing industry. Radical changes were about to take place. It was near the end of foreign factory trawlers harvesting mammoth quantities of fish off the coast, and the beginning of federal fisheries management within two hundred miles of our shores to restore fish stocks that had been seriously depleted.

Though I didn't attend annually, I went to the show every few years. Over those years I realized I had witnessed industry trends from the last quarter of the twentieth century through the beginning of the twenty-first. With exhibitors and buyers from around the world, the International Boston Seafood Show, a trade show not open to the public, proved to be my window to the fishing industry at both the local level and globally.

The history of the fishing industry in the North Atlantic dates back at least to the sixteenth century, when the Europeans' efficiency at catching fish close to home, even with what we would consider ancient technology, outstripped the supply. They sailed west in search of new stocks of fish and found enormous wealth from the North Atlantic's undersea banks. Newfoundland became the go-to place and remained so for hundreds of years. Then, when Europeans settled in New England, fishing local waters and the banks soon became an important source of prosperity for a new nation.

The Cape Cod story is intertwined with the global industry and cannot be separated from the larger picture. By the mid-eighteenth century, fishing boats were sailing from every Cape Cod port, venturing over a watery highway to Georges Bank, the Gulf of Maine, Nova Scotia, and Newfoundland. They were joined by ships from other Massachusetts ports, primarily Gloucester, Salem, and Marblehead, and ships from European countries, where, together, they harvested fish from one of the richest marine areas in the world.

The story is not just about cod either, although cod was a dominant factor in the growth of the industry. Herring and mackerel also played a valuable role. The industry is not now, nor ever was, static. A common trait among fishermen has always been the ability to adapt to changing conditions and diversify in order to survive. If one species became depleted, fishermen sought another and another and another, often traveling farther and farther away from home ports. If one fishing method no longer worked, they adopted a different one. Technology evolved as well, and with each technological advance, more fish were caught. This relentless pursuit continued until the seas could not produce fish fast enough to sustain the pressure put on them.

Ingrained in the survival of this industry is the concept that the fisheries are a common resource, free for the taking. No individual, country, company, or other entity owns them. Fishermen traditionally took whatever species they could wherever they could. Every fisherman, boat owner, or country maximized their catch by taking as much as they wanted or their craft could hold. The attitude was that if they didn't get it, the next guy would, with little regard for the size or reproductive potential of the fish.

That attitude spawned what Garrett Hardin called the "Tragedy of the Commons." In an essay published in *Science* in 1968, Hardin used an example of a cow pasture that was open to all to illustrate his point. He described a common pasture that could support a fixed number of cows, utilized by several farmers who wanted to maximize their personal gains. When one farmer decided he could increase his income by adding a cow, each of his neighboring farmers wanted to add to their income by adding an animal, too. At first, there did not seem to be an impact, so everyone added an extra cow, and then another and one more. Eventually, the pasture could not support the increased number of cows, and the farmers all lost money. Hardin concluded, "Freedom in a commons brings ruin to all."

The same situation occurs when fishing a common resource; and over years of taking, fishermen could not help but notice that fish stocks were declining.

For centuries, fishing methods remained relatively unchanged, only tweaked now and then. But profound changes occurred early in the twentieth century. Fishermen switched from using single lines and hooks dangled from the rail of a ship to long lines with thousands of hooks along the seafloor. Steam trawlers introduced before World War I could drag massive heavy nets along the bottom and inadvertently alter the habitat in the process. Compasses and sounding leads were replaced by electronic fish finders and navigational aids. Then in the mid-twentieth century, England upped the ante by launching the first factory trawler, capable of roaming all the oceans and processing thousands of tons of fish at sea.

Other nations followed suit with their own factory ships. Massive machines like those displayed at the seafood show were used to process thousands of tons of fish and flash-freeze them on board. But 1976 was approaching the turning point for the foreign fleets, and the seafood show was remarkably different a decade later.

The U.S. declared a two hundred-mile limit in 1976 to stem the tide of foreign competition, and Canada followed suit in 1977. The boundary in the U.S. was complicated by states' rights. The states controlled waters from their shores to three miles out except for Texas, western Florida, and Puerto Rico, which declared nine-mile limits. Federal jurisdiction extended all limits out to two hundred miles. The three-mile limit is complicated and an important consideration for New England and the Gulf of Maine, where cooperative management and consistency is necessary in an area shared by three American states, two Canadian provinces, and the federal governments of both nations. In 1982, the United Nations Convention on the Law of the Sea created exclusive economic zones, EEZs, where each country, called a "state" by the U.N., had "special rights regarding the exploration and use of marine resources," extending two hundred nautical miles from its coast.

By 1986, the big processing machines were still being exhibited at the seafood show, but fewer of them were displayed. Farmed shrimp, mostly from Asia and South America, and U.S.-farmed catfish were conspicuously displayed as well as Surimi, a paste made from fish but sold as a crab substitute or as an ingredient in seafood salad that had been introduced a few years earlier. A new product entered the picture: farm-raised freshwater tilapia, both domestic and foreign. Ocean fresh-fish displays were replaced by freezer showcases. Chinese companies were interspersed throughout the exhibition. Several years later, following protests at Tiananmen Square and the rise of their form of capitalism, Chinese companies dominated the show with all types of frozen seafood.

Within a few years into the twenty-first century, value-added products were displayed – portion-controlled stuffed salmon or Alaskan cod, perfectly seasoned and packaged; or fish packed in a sauce. Farmed Atlantic salmon became notable, mostly from Norway, Canada, and Maine, but Chile was back in the game as a primary supplier several years after recovering from a difficult production problem. Evidence of cod was sparse, and talk of cod in New England centered on declining resources and increasing regulations and the need to focus on alternative species and products.

By 2014, the show had shifted to the Seaport Convention Center in Boston's newly revitalized seaport district, an even larger venue. Aquaculture, an industry sector that had been steadily increasing, was more prevalent than wild harvest. Cultured oysters, clams, and mussels from many Atlantic states as well as Canadian Maritimes; farmed salmon, shrimp, catfish, and tilapia; and vacuum-packed frozen seafood from Asian companies, including farmed shellfish, fish, and shrimp, were all prominently displayed. Many products were geared to the consumer who wanted seafood but also wanted ease in preparation. Packaging became an important consideration – attractive graphics, crisp product photography, and heat-and-serve convenience – anything to elevate a product over the competition.

The big processing machines were still there, such as the Baader® Model 582, which could be operated by one or two people and was outfitted with electronic sensors to make the adjustments for different sizes of fish. It could head, gut, and fillet fish up to three feet long, including cod, haddock, pollock, Pacific cod, Alaska pollock, pink salmon, and hoki, a member of the hake family found around Australia and New Zealand, all at a mind-boggling eighty to one hundred fish per minute instead of the thirty per minute that the older model could merely fillet.

But the Baader® display at the show was much reduced in scope from decades earlier. According to Bert McBride of Baader®, "The Boston show has transformed over the years. We used to run our equipment with fish at the shows, but the costs to do so have become prohibitive. Nevertheless, we have attended the Boston show every year for decades now and will continue to do so."

The decline of wild harvested fish at the seafood show was mirrored in the Northeast's fishing communities. Garrett Hardin's tragedy of the commons proved all too true, and as demand for the valuable seafood outstripped supply, fisheries became a highly regulated industry. As the wild harvest waned with declining stocks of fish, farming the seas through aquaculture emerged as another way to provide that cherished seafood.

But harvesting from the commons has always been a natural part of human existence and is ingrained in our behavior. Those who harvest from the fishery commons are generally unwilling to voluntarily alter their behavior to save fish because they know it may ultimately mean giving up their way of life. Regulatory oversights curb this behavior, but regulations have not always been effective as exemplified by the difficult recovery of Northeast fisheries, despite ever-increasing restrictions. And each time a regulation that attempts to limit an activity that has an economic or cultural impact is proposed, there is a backlash. Particularly distressing to many are the regulations on ground-fish, especially cod.

The Cape did not escape the tragedy of the commons. It took several centuries between 1602, when Bartholomew Gosnold named the peninsula Cape Cod, and the twentieth century when the decline of the fish became painfully obvious. The fishing community on the Cape was caught in a vise with external forces turning the screw. Foreign fleets of factory trawlers were followed by large, technically advanced and highly capitalized U.S. boats, followed by further declining harvests, followed by severely restrictive U.S. regulations, followed by total protection of marine mammals, and finally followed by dramatic changes in the ecology. On land, working waterfronts morphed into tourist attractions and property prices excluded many working fishermen from owning homes. The forces conspired to decimate a once-proud industry.

While all this was going on, the Cape Cod fishing industry did what it has always done – adapt and diversify. In the process, some continued to fish commercially while others switched to lobstering, or began running charters for recreational fishermen, or established eco-tours of the surrounding waters. Others gave up the sea and began catering to both seasonal and year-round landside customers and went fishing recreationally when they could.

The situation is complex. There are many players vying for a piece of the common resource pie, for an independent way of life, for keeping a centuries-old tradition alive, and for making a living on the water. The story continues to unfold, along with the human drama of not always being able to have what we want when we want it all.

## Chapter 10
# The Rise of King Cod

The men arrived at Orleans' Snow Shore landing in the pre-dawn blackness and donned their foul weather gear. The stiff, non-breathable, bright yellow and dull orange waterproof jackets and pants and heavy rubber knee boots protected them against rough conditions and the cold sea. Accustomed to working in low light from years of experience, they backed their trucks onto the hard-packed sand and gravel to unload the well-used gear they would need for their day on the water. It was still raw on this early April morning in the late 1970s, but the forecast called for a lull between weather systems, and they needed the cash after a lean winter. Cod were plentiful, and a good catch would help to pay some bills. In several months, most of them would switch to setting traps for the more lucrative lobsters crawling along the back-side of the Cape, but at this time of year, cod were available. Flexibility was essential to being a successful fisherman here.

The inlet was not too rough, with an ebbing tide and gentle wind, not the pounding northeast gales. The fishermen could sneak out without too much trouble, although for Nauset Inlet, "not too much trouble" would scare most normal people to death. Everyone who fished out of Nauset had a tale to tell about a near miss or knew someone who had gotten into trouble – serious trouble – at one time or another.

There was no marina or loading dock at the landing, just room for eight trucks to park, and a black-topped ramp that ended on an intertidal strip of flat, hard-packed sand, stones, and shells. Even at high tide the water was not deep enough to efficiently launch a boat from the ramp, so the only people who launched there were those with small skiffs easily pushed off a trailer. On the left of the ramp, a historical plaque marked the spot where Samuel de Champlain had entered the harbor in 1605. Beyond the plaque, a natural shoreline of beach and salt marsh led to the closest house about five hundred yards away. In sharp contrast, on the right side of the ramp a massive manmade rock revetment extended for at least five hundred feet, protecting the shorefront home and its boathouse from the relentless winter storms. Only the boathouse was visible from the ramp.

**Fisherman at Nauset Harbor with no commercial fishing dock.**
**(Photo by author)**

At low tide the fishermen backed their trucks down to the water to unload their gear. Their thirty-five to forty-five-foot boats were moored in the narrow channel, a mere hundred or so feet wide. Beyond the boats, shoal water led to expansive and exposed sand bars, their sizes and shapes constantly changing. The landing and the harbor were not what one might call commercial-fishing-friendly, but they were the closest points of access to the inlet and offshore fishing. A lot of steps that took precious time were required to get ready, but the men put up with the inconvenience and ordeal. And when they returned, they would reverse the process – except that, hopefully, their fish boxes would be brimming with gutted, chilled cod.

First, the fishermen dragged a dinghy anchored on the beach to the water and waded out just far enough to float it, careful not to go out so far that the numbingly cold water would pour over the tops of their boots. They rowed or paddled to a moored eighteen-foot skiff and left the dinghy on the mooring. They motored the skiff back to the landing and loaded it with jerry cans of fuel, fish boxes with ice, and the rest of their equipment; then they parked their truck, walked back to their skiff, and headed out to their big boat, moored fifty yards offshore in what passed for deep water – "deep" being no more than eight feet at low tide. Once the larger boat was loaded, the men tied the skiff to the mooring and gingerly threaded their way in the slightly brightening light through the narrow convoluted channel to the inlet.

They threaded their way through the channel aided by the unofficial buoys they had made, topped with aluminum radar reflectors to guide them. They had installed them themselves and repositioned the buoys whenever the sands shifted after a storm and the channel changed location.

Nauset Harbor is volatile. Sand bars and salt marshes separated by fast-moving, shallow, shifting channels define much of the harbor. But if the harbor is volatile, the inlet is treacherous. It is a narrow opening through a barrier beach with a seven-knot current, constantly shifting sands, shallow bars on the outside, and breaking surf. Often the configuration of the bars means that a boat has to run parallel to the beach and broadside to the surf – a highly dangerous maneuver. Negotiating these risks takes skill and guts. Champlain had named it well in 1605: Mallebarre, "bad bar."

Once safely outside the inlet, the men headed offshore for a couple of miles, jigging for cod. A heavy, shiny metal lure, or jig, was tied to thick monofilament line, often with brightly colored "teasers" attached above the jig. The line was wound around a wooden or plastic spool, and they lowered the jig to the bottom and bounced it a time or two. If there were plenty of cod, that was all it took to hook a fish of ten to twenty-five pounds. Then, time after time, the fishermen hauled the fish hand over hand to the boat where the cod were gilled, gutted and tossed into the fish box to be iced down.

Jigging, often described as "a jerk on one end waiting for a jerk on the other," is a fishing technique not much changed for centuries. If the fish are plentiful, fishing requires no more than brawn and stamina. But when there are few fish, fishing requires more than sheer strength. It takes knowledge and experience, various techniques, finesse, baited hooks, fish finders, and other modern equipment.

Heading home was more a matter of timing than the number of fish that were caught because the fishermen had to work the tides inside the harbor, which was not negotiable at low water. As they approached the inlet, the fishermen scoped out the surf, sometimes lingering slightly offshore, timing their reentry just right to avoid real danger. It was one thing to go out the inlet, knowing you had to maneuver through the surf and hoping that a big wave didn't crash down on top of you. It was quite another to run before a big wave, looking aft and seeing a wall of water over the stern, knowing if you did not run fast enough, or if the engine faltered at just the wrong moment and the wave caught up to you, you had no control. It would swamp the boat in a flash.

As dangerous as it is, Nauset Inlet has been plied by fishermen for centuries. Before Europeans found these shores, the Nausets, the Native Americans for whom the inlet was named, paddled in and out of its risky waters for countless generations. Their fishing was not a commercial enterprise; the fish off the coast enabled them to survive. They negotiated the surf, shoals, and sand in their mishoons with skill and a deep understanding of the complex world and the secrets of the sea. As Chief Earl Mills explained, the Nausets also gave thanks for the gifts of the ocean, the fish and whales they relied on to sustain them.

Like the twentieth-century fishermen, the Nausets' day began early. Imagine a summer dawn at Nauset Inlet in the sixteenth century: the cloak of darkness gently lifting, revealing a brightening sky to the east, the palette changing from dark purple to mauve to pink to yellow until finally the golden sun emerges from the ocean's horizon, rising in the sky, the start of a glorious day. From their small wetus, the summer camps situated on the bluffs surrounding the harbor, the Nausets looked out across the glass-calm embayment to the sea beyond. They were the only people on this far-eastern arm of sand jutting into the ocean.

The strong young men pushed their dugout canoes from the shore and paddled swiftly, cutting across the harbor. Soon reaching the mouth of the inlet, one of only two breaks in an otherwise continuous thirty-plus-mile line of sand, they maneuvered their canoes through this treacherous passage as they had countless times before, embracing the knowledge handed down orally in story and song. When the inlet was kind, fishing was less dangerous, although danger was no stranger to them. At less opportune times, breaking waves could tower above their canoe, the ocean a cauldron of bubbling froth and pent-up power.

They did not need to paddle far from the land before they saw signs of fish. Gulls and terns, vigilantly searching for food, circled above them, diving into the water for a bite. Birds "working" are a sign of fish nearby. The men knew that in the summer cod were abundant a mile or so offshore in the colder, deeper water while striped bass prefer the warmer water closer to shore. Both types of fish fed on the plentiful sand lance that were everywhere, drawing both birds and larger fish to the region.

The men filled their canoe with fresh fish sufficient to feed the village for a few days, or they might smoke or dry some of the fish for another time.

Skilled Wampanoag people and their predecessors were fishing Cape waters for at least ten thousand years. Whether the Vikings reached the Massachusetts coast during the last millennium is still speculation. But they, and later the Basques, were frequent visitors to the general area following the richness of the banks along the way, and fishing mostly for cod. The trouble the industry is in now can be traced at least as far back as medieval times in Europe as commercial fishing expanded and attitudes toward fish and fishing, harvesting the sea's riches for economic prosperity, became paramount.

Rumors of abundant cod to the west across the sea circulated among European captains and fishermen for decades, even before the Italian Giovanni Caboto, better known as John Cabot, made his historic 1497 exploratory survey of the New World's shores. The Basques had been bringing home whale oil and fish from the New World long before Cabot, never divulging where they went, where they caught their catch, or what riches they saw.

Reaching Newfoundland was a treacherous undertaking even with ships one hundred feet long that were among the most advanced vessels of their day. But fishing was far more profitable, even for the lowliest man on board, than struggling to survive ashore. The Basques traveled across the ocean to hunt whales, but the sheer volume of fish they found along the North American coast was astounding. Many Basque boats turned exclusively to fishing. With luck and a mild crossing, a five-week voyage was rewarded by easy fishing — not that anything about fishing was truly easy.

It was not only the blending of cold and warm currents that made Newfoundland an extraordinarily productive fishing area. Shallow banks that dropped off into deep canyons were the other factor. Warm and cold, deep and shallow magically result in the ocean's richest habitats and an incredible diversity of species, ranging from microscopic plankton to giant whales found in dozens of places in the northwest Atlantic.

Newfoundland was a lonely, forbidding place, shaped by the water and wind. Stunted evergreens sparsely covered steep rocky hills at the edge of the sea, alternating with coastal cliffs of tundra worn flat by the elements. The deep fjords lining the crenulated coast formed natural harbors, although few were shallow enough for anchorages. Where anchorage was found, the Basques erected drying flakes on shore, wooden racks necessary for dry-curing fish.

For decades, the Basques claimed Newfoundland's vast harbors for themselves until word of Cabot's discovery of the rich resources off the

New World's coast reached Europe's seafaring countries.

The fisherman's code was to remain closed-mouthed about where they caught fish, a code especially true for captains. But grog could loosen the lips of seamen with a fantastic tale to tell. Fish were so plentiful they could be caught by merely lowering a weighted basket into the water, or so it was reported. It was practically impossible to keep fish stories like that a secret for long. Taking advantage, Spanish, Dutch, French, and eventually English financiers underwrote the dangerous journeys to the greatest of cod fisheries.

Why would fleets of boats from four European countries, risk so much to venture so far during the Middle Ages? There were fish closer to home, but even during the sixteenth century demand exceeding supply for plentiful protein food was a problem. Overfishing is not a new concept.

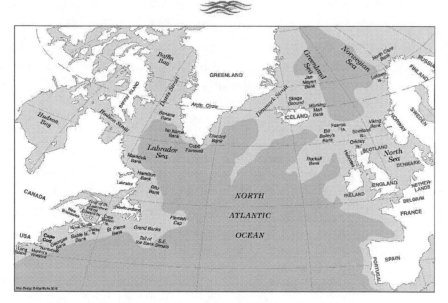

**Map of North Atlantic with specific banks. (Mapworks)**

Almost everywhere the Norsemen explored during their transoceanic voyages – the North Atlantic, North Sea and Baltic – from 1050 to 1500, they discovered waters teeming with fish, whether close to home at the Lofoten Islands off their own shores or far away in the New World. Viking sagas place them on the shores of the North American continent around

AD 1000 at L'Anse aux Meadows, the northern tip of a long peninsula on the western flank of Newfoundland, where the Norse settled briefly.

While cod was a vital catch, it was actually herring that served as the critical link, providing food for the masses as well as the fish further up the food web, including cod. Herring was the most important fish consumed in the coastal communities of northern countries where fishing was done close to shore and the fish could be eaten fresh. As Christianity spread across Europe, the demand for fish steadily increased as inadequate supplies of meat combined with Catholic dogma that required meatless meals on holy and fasting days, which at one time totaled almost half of the year. The sheer abundance of herring in the Baltic and North Sea provided a religiously-sanctioned source of protein for the masses on the meatless holy days, explained Brian Fagan in his 2006 book *Fish on Fridays*.

Fish merchants enjoyed high profits, and fishing, especially for the massive schools of herring, became a lucrative industry. However, herring had a short shelf life. A cheap method of fish preservation was essential, but herring is an oily fish. It did not take well to salting, smoking, or drying, the common preservation methods at the time. Most peasants, fishermen, and military men survived on heavily salted herring, a dry jerky-like product that was often non-palatable, tasteless, or downright rancid, Fagan stated.

Salt, evaporated in large vats, was essential for preserving fish and readily accessible in Spain, Portugal, and France, but the climate in England and northern Europe was not conducive for massive salt evaporation operations. Salt became a commodity traded between "have" and "have-not" nations, a source of wealth and power. At the time, national navies did not exist to protect shipping interests. Instead, trade merchants banded together, forming the Hanseatic League in 1100, a powerful and influential cartel centered in Germany which affected all fish-producing nations. The league governed everything from setting prices to shipbuilding to marketing large supplies of salt obtained from merchants in southern Europe, explained Fagan as well as Mark Kurlansky in *Cod* in 1997.

Cod, and to a lesser degree its relatives – haddock, pollock, ling, and hake – were extremely abundant and more palatable by far than herring. The cod's firm, white, non-oily flesh took well to preservation but also required vast amounts of salt to be barreled in brine and dried. However, the climate around the Lofoten Islands, above the Arctic Circle, was not only a perfect cold-water habitat for cod. It was conducive for air drying without salt, preserving the fish for lengthy amounts of time – up to five years if kept dry – without losing quality. Islanders erected large air-drying racks called flakes for gutted and split fish, a centuries-old tradition still prevalent today, according to multiple reports.

The popularity of stockfish, dried and beheaded unsalted cod, underpinned the economy of northern Norway. Cod fishing was the primary occupation of the Lofoten islanders, who sailed north of the Arctic Circle under oar or sail in open boats, catching fish one by one, hand over hand, in all types of weather. Yet the fishermen barely eked out a living despite their arduous work. It was miserly merchants, the middlemen underwriting the fishing vessels and large transport ships, who made their wealth at the fishermen's expense, a continuing industry issue.

Fishing was a fundamental industry throughout the British Isles, too. Harvested to the brink, their herring fishery collapsed late in the fourteenth century. The collapse curiously coincided with the onset of the "Little Ice Age," a climatic event that plunged northern Europe into a frigid icebox from roughly 1250 to 1850, wrote Fagan. When the all-important herring became less of a sure thing, merchants sought alternatives to satisfy religious and population requirements. Cod rose as the alternative of choice, preserved by the Norse method of unsalted air drying. Cod fishing intensified all around the British Isles. Eventually, demand outstripped the supply, landings decreased as overfishing continued, and the Britons were forced to fish farther from their home ports to meet demands, at greater cost and danger, stated Fagan and Callum Roberts in his 2007 book, *An Unnatural History of the Sea.*

By the fifteenth century, the English were sailing to Iceland. They fished off the southwest coast where stocks were plentiful, going ashore to salt and barrel the catch. The Icelandic people, who made short daytrips to fish close to their home shores, banned overwintering by the foreigners. But for English mariners, a homeward journey without a full catch was inconceivable. Fishing around Iceland increased dramatically as three other nations vied with the English for supremacy – the Norse, Dutch, and Bretons, Fagan and Roberts related.

Cod landings skyrocketed, but cod was still second to herring. There were enormous profits to be had – but also huge losses – and no shortage of men willing to risk everything to be a part of it. With the fishing pressure exerted by the four major nations, the Icelandic fishery faltered, became unsustainable, and declined rapidly, Fagan added.

The demand for fish remained high, and those who could supply fish, even if they had to voyage farther away, profited exorbitantly. Going farther away meant Newfoundland and the Grand Banks. Fishing off Newfoundland increased dramatically and was intense by the middle of the sixteenth century. Finally, around 1570, the English joined the Newfoundland cod fishery. They transformed the air-drying method by adding salt to the process, turning their catch into a cured, dried salt-cod product that is still sought after today in some locales. This preservation method and the fish's abundance set the stage for cod to become king of the Atlantic. More importantly, the English colonization of Newfoundland laid the groundwork for colonization in Jamestown, Virginia, and New England, launching a new nation.

Thus, by the sixteenth century, fully a century before the founding of Jamestown or the splash of the *Mayflower's* anchor, Europeans were fishing in waters off the New World every summer.

And the long-at-sea Basque fishermen, who had secretly fished the prolific banks off Newfoundland for generations, headed home. They had caught a boatload of cod, a fish destined to change the world, from a place that was no longer a secret.

Several hundred miles southwest of Newfoundland, the Nausets glided their dugout canoes back into Nauset Harbor with cod for their village. The world as they knew it was about to change as well. They would never be the same.

And the fishermen of the late twentieth century came back safely to Nauset Harbor with fish boxes of iced cod. Their world, too, was about to change as the fish that was once so plentiful ran out and their historic, time-honored industry became unrecognizable.

# Chapter 11

# Warm and Cold, Deep and Shallow

Mid-summer throngs of people stroll on the sidewalk and along Provincetown's Commercial Street while cars inch their way through the town. Tourists watch the colorful local characters while the residents watch the visitors taking in the sights and sounds of the tourist businesses – the hotdog stands, bars, cafes, and restaurants; salt water taffy, T-shirt, jewelry, and souvenir shops; art galleries, small inns, and B&Bs. A walk along the shore reveals more restaurants, with seating for "alfresco" waterfront dining and wooden decks attached to weather-worn wooden houses, so close to one another that few of them have side yards.

It can take a bit of imagination for tourists to conjure up images of a time past, when the sandy shoreline was covered with long latticework wooden racks, called fish flakes, laden with thousands of split-cod slabs drying in the sun. Dried cod enhanced the community's ability to sell fish to far-off markets and to have high-quality protein for their own needs. They did not develop the method on their own, though. In the early eighteenth century, Cape fishermen, sailing hundreds of miles to Newfoundland, brought back more than fish. They returned home with the knowhow to preserve the fish, learned from northern fishermen who had been drying cod in this manner for centuries. In the eighteenth and nineteenth centuries, Commercial Street was not filled with tourists and gawkers and the stores and shops catering to them. It was filled with fishermen and people working the fish flakes and with businesses catering to the fishing industry.

Newfoundland had perfected the art of "making fish," as it was called.

It was a source of pride for "Newfies" to make a high-quality product that was then shipped to the discerning European market. For generation after generation, summer in the coastal villages meant "making fish." What appears at first glance to be a relatively simple, unsophisticated process was actually a series of complex tasks in which skill, tradition, stamina, and knowledge all played a part. When added up, it was an exhausting, laborious community effort – men, women, and children working nearly around the clock, with little rest, living a mere step above subsistence.

Fish arrived every day at the "stages" – sometimes twice a day during a prosperous season. The fish were headed, gutted, and split. Bucket after heavy bucket of water was lifted high above the sea to the docks to thoroughly wash the split fish. Then the fish were hauled in wagons to the salter, usually a woman. They were salted and packed systematically in barrels, according to weight with big fish on the bottom to absorb more brine and smaller fish on top. They remained in the barrels for three to four days. The next day, more fish arrived, and the tasks were repeated.

**Stages at Newfoundland cod processing village. (Credit: Mark Ferguson from "Hard Racket for a Living - Making Light-Salted Fish on the East Coast of Newfoundland")**

When the first batch was fully brined, the brine was washed off with more buckets of seawater and the cod was spread on the wooden latticework flakes. That evening, those fish were removed from the flakes and methodically piled for the night. The next morning, the batch of cod was spread on the flakes again. After washing, the second batch was also spread on the flakes. When ready, the first batch were piled into fagots, round piles where fish were layered skin-side out, for a few days and then spread on the flakes again. The shore skipper, also a woman, tracked the batches of fish going through this painstaking process. Each day, the fish required a particular step in the constantly rotating process until, after one to two months, depending on weather and other factors, the final product could be shipped.

**Newfoundland fish flake. (Mark Ferguson)**

Putrid odors filled the area. At low tide, the offal from the thousands of pounds of heads and guts dumped into the water would cook in the summer sun with the stench of rotting fish until the tide rose and washed it offshore. At the salting stage, the used fish-flavored brine was emptied onto the rock ledges where it ran down to the sea as the fish were washed.

The cod piled in various configurations and drying on the flakes sent a different scent wafting through the air. The process must have taken the uninitiated quite a while to grow accustomed to the smell.

The Newfoundland lifestyle combined communal and personal traits that developed the character of the people as industrious, strong, resilient, intrepid, and frugal, with an ability to work together for the common goals of economic need and personal survival. The villagers' strong work ethic defined the fabric of the community, where every individual took part in the process of making fish. They were, figuratively speaking, all in the same boat together wrote Mark Ferguson in "Hard Racket for a Living" in 1997.

**Fish flakes in Provincetown. (Courtesy of Cape Cod National Seashore)**

Provincetown was also a fishing community. Catching, processing, and marketing fish is what they did. However, contrary to the Newfoundland manner, in which the splitter was extremely careful to make precise cuts to ensure a high-quality product, in Provincetown the fish were split quickly to create a butterflied effect and spread on the flakes. The fish were also salted in barrels but not in a carefully layered manner. This less meticulous technique did not meet the same high standards as Newfoundland's

115

because instead of going to Europe, Provincetown's salted dried fish were shipped to the Caribbean to feed slaves working on sugar plantations. However, if the Provincetown process was not quite as complex, it was still labor-intensive and produced the same smells.

≈≈≈

For a time, from the late eighteenth to the mid-nineteenth centuries, the Cape's shores were the site of another important industry – making salt. Preserving fish required lots of salt. Initially, colonists boiled ocean water in large vats to distill the salt. The amount of wood needed to keep the fires going caused further deforestation, with up to two cords of wood required to boil four hundred gallons of seawater to produce one bushel of salt, explained William P. Quinn in *The Saltworks of Cape Cod* in 1993.

In 1776, John Sears, an enterprising Cape Cod boat builder and captain from East Dennis, constructed a vat to evaporate water, producing eight bushels of salt during his first season. The next year, he caulked the seams and more than tripled his output to thirty bushels. Sears built a rolling roof that could be opened to allow for evaporation or closed to keep out rain and dew, and a windmill to pump the water. The industry took off.

Salt works were ingenious contraptions. The salt makers built large wooden vats, ten to sixteen feet square and nine to twelve inches deep, in descending steps along the shore. Windmills provided the power to pump the seawater through a system of wooden pipes to the highest vat for a three-step sequence, with gravity dropping the water from one step down to the next, removing the impurities.

In the highest vat, unwanted plants and animals dropped to the bottom to be removed. When enough water had been evaporated from this vat, concentrating the salt so much that no further marine life could grow, the "settling" process was done. The salty water was then drained to the intermediate vat, the "pickle room," where lime and gypsum (calcium salts) precipitated to the bottom. As crystals of sodium chloride, or table salt, formed, the water flowed on to the final step, the "salt room." There, the white crystals increased and were collected for sale.

116

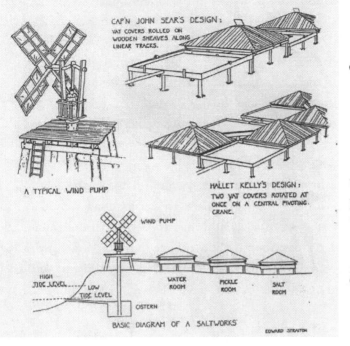

**Diagram of salt works. Salt was evaporated in large vats built in three-tier steps. Water was pumped to highest vat powered by a windmill and flowed by gravity from one vat to the next. Roofs were closed in rain.**

It still took nearly four hundred gallons of seawater to produce one bushel of salt, using solar energy and time – three to six weeks. But the season was long, from March or April to October or November. Managing the salt works took a village. Schools were reportedly closed at the first sign of rain so that the students could rush to help close the roofs over the vats. This was a successful, lucrative enterprise. Each individual salt work produced, on average, 250 bushels of salt a season.

Nothing was wasted. Yankee ingenuity increased the yield of the salt works by making use of the complex chemistry of seawater to produce other products: Glauber salt, used to produce sodium carbonate, or washing soda, and to tan leather and dye cloth; fertilizer; and "bittern water" or simply "bittern," from which was produced Epsom salt (magnesium sulfate). Boiling this with "pearl ash" from wood ashes produced magnesia (magnesium carbonate). It all generated greater profits.

The industry expanded quickly. Early nineteenth-century maps of every town show the locations of salt works and the windmills that brought the water to the works. In its heyday of 1837, 658,000 bushels of salt were harvested from 715 establishments all over the Cape. Yarmouth led with 365,000 bushels alone from fifty-two salt works. Some people called the industry "the lazy man's gold mine." Little investment or labor was required aside from building the vats, shoveling damp salt to a drying room, and adjusting the roof. Mother Nature provided the wherewithal – wind produced the power, gravity moved the water from one vat to another, and the sun evaporated the water.

But while the industry grew quickly, it was relatively short-lived. Salt deposits were discovered in New York, and the Erie Canal opened in 1825, allowing for easy transport of the salt. Nevertheless, with the value-added products, salt making continued on the Cape until the 1880s. Today, however, nothing remains of this once-huge industry beyond photos of the later structures and the occasional discovery of a salt-preserved wooden building.

The bounty of the sea was recognized early on. In 1602, British explorer Bartholomew Gosnold and John Brereton, the chronicler who sailed with him, explored the Massachusetts coast. They found an arm of sand surrounded by water that Gosnold named "Cape Cod" for the plentiful fish they saw. Brereton wrote:

> "...we returned to our ship where, in five or six hours absence, we had pestered our ship so with Cod fish that we threw numbers of them over-boord againe ... the places where we took these Cods (and might in a few daies have laden our ship) were but in seven faddom water, and within lesse than a league of shore; where in Newfoundland, the fish in fortie or fiftie fadom water and farre off."

When the *Mayflower* arrived in Provincetown, the crew and colonists must have known about the bounty of the region, even though they were originally headed for the Hudson River. Less effusive in their writings than those of earlier explorers, the colonists recounted what they had seen in *A*

*Relation or Journal of the Beginnings and Proceedings of English Plantations Settled at Plimoth in New England*, better known now as *Mourt's Relation*, written in 1622. At that time, whales were the object of their awe. They did not see cod when they arrived in November but wrote that they suspected the fish were plentiful in their season. It was not until early January, during a storm, that they began to catch cod and seals.

An agrarian group, the Pilgrims were ill-equipped for harvesting gifts from the sea, yet it did not take them long after arriving in Plymouth to attempt to rectify that situation. They learned the arts of whaling and fishing from the Wampanoags.

Their novice efforts were paltry compared with those of settlers farther north in the Massachusetts Bay Colony, an area that stretched to Maine. Fishing figured prominently in that settlement's plan to repay the English company that sold them the land patent to establish their colony, although the company had no legal right to the land. The deep harbors at Cape Ann, especially in Salem and Gloucester, became primary ports. They were close enough to Boston to take advantage of the city's merchant trade and transportation infrastructure as it developed. Gloucester remains one of the largest commercial fishing ports in Massachusetts today.

Legislators, often merchants and investors themselves, established a commission to manage the burgeoning fish trade, enacting its first regulation in 1639, stating as its purpose, "for the further encouragement of men to set upon fishing." The order granted tax exemption for seven years for fishermen, processors, and transporters; an exemption from military training; and a prohibition on using cod or bass for "manuring" – fertilizer, explained Raymond McFarland in *A History of New England Fisheries* in 1911. The Plymouth General Court recognized the special significance of the resources at the colony's backdoor, and fishing, held in high regard, contributed to the stability of the community and its growth, not just profits. In 1641, the general court ordered that space along the shore be set aside for cod processing, flakes, and stages; and granted to the owner of every fishing boat four acres of valuable upland, safe from flooding, McFarland reported. It was a substantial reward to boost personal status. Such early recognition of the intrinsic value of fishermen and shipbuilders and the efforts to encourage fishing as a valuable commodity not to be wasted, speaks volumes.

The general court recognized the benefit to future landings. It astutely enacted early laws to protect fish lifecycles that would be called environmentally sensitive today. In 1668, fishermen were forbidden from taking cod, hake, pollock, and haddock during the spawning periods of those fish, in the first months of winter. Likewise, it was illegal to take mackerel before July, wrote Jeffrey Bolster in 2012, *The Mortal Sea: Fishing the Atlantic in the Age of Sail*, as well as McFarland. Even though many members of the general court were merchants and stood to gain from the fisheries, twenty-first century readers may find it surprising that the court enacted legislation as a conservation measure to manage fisheries in the mid-seventeenth century. Surprising or not, their actions recognized ecological principles and the importance of lifecycles.

New England fishermen enjoyed the marketplace edge, too, with decreased competition from English ships. The British Civil War of 1642 drastically reduced the British fleet, and American ships filled the void. The best of American fish were sold in European ports from Calais and Bilbao to Madeira. In Jamaica and Barbados, the lower quality fish was exchanged for wine, sugar, and molasses, much-needed staples back home and in the European economy. Return trips across the Atlantic brought valuable commodities the colonists could not manufacture such as china and cotton cloth.

Shipbuilding was essential to carry the cargoes of cured fish products to markets in Europe and the West Indies. The two industries were an essential link in what became known as the "triangle trade." That expression later took on sinister connotations with the triangular slave trade. The success in fishing and shipbuilding in the Massachusetts Bay Colony laid the groundwork for other commercial success and fostered jealousy in the mother country, sowing the seeds of the rebellion and political upheaval that followed.

By the last half of the seventeenth century, fishing had become the most important industry in Massachusetts, an engine of prosperity that formed the basis of the fledgling Atlantic colonies' economy. One hundred years after Plymouth was settled, Cape fishermen were sailing to Nova Scotia and the Grand Banks off Newfoundland, curing their catch on beaches along the way or coming back to the Cape to dry the fish before heading to the West Indies. So wrote John T. Cumbler in *Cape Cod: An Environmental History of a Fragile Ecosystem* in 2014.

Although the local salt industry ended, fishing continued, remaining crucial to the Cape for the next three hundred years. Underscoring the significance of cod to the Commonwealth of Massachusetts, a wooden model of a cod, named "The Sacred Cod," has hung from the ceiling of the House of Representatives chamber in the State House in Boston since 1784. The weathervane perched atop the Cape Cod Five Cents Savings Bank, a local institution since 1855, has been a prominent fixture influenced by the winds and changing directions through the years. It is a clever combination of a cod and the shape of Cape Cod, with a cod representing the land mass from Falmouth to Orleans merging with a map of the Outer Cape from Eastham to Provincetown. As the logo of such a long-standing institution, it symbolizes the importance of cod to the Cape.

The cod was not just critical to shaping the Cape Cod economy, though. It represented a way of life. Yet, it has also been the basis of heartache. Today, it is almost impossible to imagine a sea so full of the bottom-dwelling cod that "inexhaustible" was often used to describe its abundance. Now, cod has dwindled precipitously, and "collapse" is more often used to appropriately describe its status.

**Weathervane atop all branches of Cape Cod Five Cents Savings Bank featuring body of cod as shape of Cape Cod. (Photo by Kim Tellert)**

The tremendous amount of western Atlantic cod, first reported by Cabot, fed the hunger for fish in Europe and North America for half a millennium. The schools that thrived off Newfoundland, the earliest hub of North American cod fishing, seemed limitless. Farther southwest were other areas just as bountiful. Taken together, the banks off Newfoundland, Nova Scotia, and the Gulf of Maine, as well as Georges Bank, became among the most productive fishing grounds in the world, aiding the economies of many European and North American nations for centuries. This lucrative fishing was made possible because of the hydrography – the deep and shallow bathymetry combined with warm and cold currents – supporting marine populations' needs to feed, grow, and reproduce. Today, scientists call this a Large Marine Ecosystem, areas of ocean covering eighty thousand square miles or more in coastal waters adjacent to the continents where primary productivity is generally higher than in the open ocean.

The Gulf of Maine, part of the larger LME, features that combination of shallow and deep, warm and cold, resulting in spectacularly productive habitat. Fishermen throughout the region worked in Massachusetts Bay and beyond, into the gulf where fertile fishing grounds were close enough that, as the coastal populations increased, every marsh creek or shallow cove or protected inlet became a port of sorts for the local fishermen.

The Gulf of Maine is a watery gem. Covering thirty-six thousand square miles, the crenulated shoreline borders seventy-five hundred miles of Nova Scotia, New Brunswick, Maine, New Hampshire, and Massachusetts, and includes Georges Bank and Browns Bank. Those raised underwater plateaus define the outer seaward edge of the gulf and form the seaward edge of the continental shelf, beyond which the ocean topography descends thousands of feet. Cold, nutrient-rich water enters the gulf through the Northeast Channel between Georges and Brown Banks and flows in a counterclockwise direction. Stellwagen Bank and Cashes Ledge are two undersea islands in the interior of the Gulf, and several deep basins – Jordan, Wilkinson, and Georges, each over six hundred feet deep – enhance the bathymetry where the water flows over the shallow banks and ledges and through the deep basins. The area also includes the Bay of Fundy, home to the highest tides in the world, with a tidal range of nearly fifty feet. The Gulf of Maine's sixty-nine thousand-square-mile watershed reaches as far inland as Quebec, adding a freshwater ingredient that mixes with the salty ocean water to allow estuarine species to flourish as well, further magnifying the biological splendor. Such rivers are important for anadromous species – fish that spawn in fresh water and spend their adulthood in the sea. They include river herring such as alewives, blueback herring, and shad, plus catadromous species – fish born at sea that spend part of their lives in fresh water, such as eels.

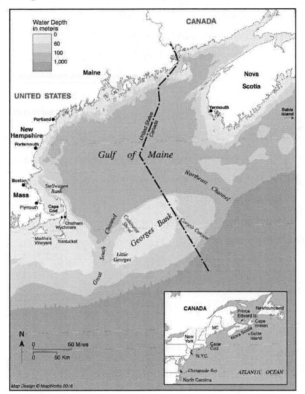

**Map of Gulf of Maine. (Mapworks)**

Two hundred twenty fish species alone utilize these varied environments for some part of their life histories, wrote Henry Bigelow and William Schroeder in *Fishes of the Gulf of Maine* in 1953. That makes the Gulf of Maine one of the richest, most productive, diverse, and unique marine communities in the world. Stellwagen Bank, where cold northern and warm southern currents meet and swirl (together with seasonal variations), is home to the 842-square-mile Gerry E. Studds Stellwagen Bank National Marine Sanctuary, established in 1992 as the first marine sanctuary in New England. Named for Congressman Gerry Studds, a longtime fisheries advocate, and located twenty-five miles east of Boston, the sanctuary is an undersea bank that stretches from five miles north of Provincetown to five miles east of Gloucester. Within its boundary, measuring nineteen miles north to south and six miles across, the sanctuary's high productivity sustains over 575 known species, including eighty species of fish, twenty-two kinds of marine mammals, thirty-four species of foraging seabirds, and numerous invertebrates such as lobsters, squid, and sea scallops.

With some exceptions – Provincetown, Wellfleet, Chatham, and Yarmouth – the Cape did not have the deep, protected harbors to accommodate the larger schooners and other vessels needed for long-distance offshore fishing that could withstand the wilder conditions on Canada's Grand Banks. But the Gulf of Maine's calmer waters did not require those larger vessels. Cape fishermen, with their smaller boats that could be kept in the local shallow creeks and coves, could fish in the gulf and in the productive waters surrounding the Cape. Fishing from these small ports became known as "inshore fishing." While there is no clear distinction between the terms "inshore" and "offshore," inshore is close to the shore and most often now refers to "day trips," although in fishing terms, "day trips" can sometimes be two or three days.

The Cape may not have had deep ports, but what it did have were skilled sailors who were sought after as captains and crew – and for war. Both the American Revolution and War of 1812 interrupted commerce, from 1775 to 1783 and again from 1812 until 1815. The wars required able-bodied seamen, and New England fishermen were the best. England blockaded harbors such as Boston, disrupted the flow of goods by refusing access to American ships, and employed privateers to destroy vessels in acts of war, disrupting the triangle trade. Many of the wooden fishing boats that made up the remaining fleet were abandoned during the several years that trade was interrupted. The boats slipped into disrepair at first and were finally destroyed. The same was true of whaling vessels. Consequently, the role of fishing diminished for a time.

It is not clear whether the smaller boats used mostly for inshore fishing from the numerous Cape creeks were destroyed during those times of conflict. What is apparent is that with many boats destroyed and a denuded landscape difficult to farm, the Cape had become a difficult place to live.

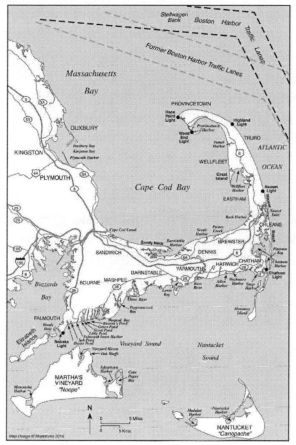

**Map of Cape Cod with small harbors. (Mapworks)**

Fishing endured, though. Those who remained eked out a living as best they could. Once again, they turned to the sea, the resource they knew so well. Though rebuilding the fleets took years, it was a worthwhile investment because the maritime industries came back stronger than ever.

Through the nineteenth century, the best estimate is that three-quarters of the Cape's able-bodied men made a living from the sea in some capacity – by fishing, whaling or in trade.

Provincetown became the center of fishing because it was close to Stellwagen Bank and other parts of the Gulf of Maine as well as offshore grounds on the "back side" of the Cape and relatively close to George's Bank. Its deep harbor could accommodate larger vessels, and the shore was equally well-suited to curing and drying cod to preserve the fish.

Provincetown was not the only port to erect fish flakes. Wellfleet and the south-side beaches had them too. Fish flakes were even constructed on the shores of Nauset Harbor and Chatham, indicating that despite the difficulty of navigating the treacherous inlets with their constantly shifting bars, strong tidal currents, and pounding surf, the close proximity to fish was a powerful lure.

Paradoxically, even with the extensive fish flakes, there is little evidence that Cape Cod shared in the great cod fishing bounty that the rest of Massachusetts experienced during the early colonial period or in the heyday of the nineteenth century. Fishing methods changed, and as important as cod was, with the exception of those few towns named above, the Cape simply did not have the harbors or infrastructure to become an offshore fishing hub. Numerous histories stress the importance of fishing, noting the great percentage of the population that earned a living on the water. Yet, despite its name and the historical records, the documented numbers of fish taken from the sea do not substantiate Cape Cod as a premier location for cod fishing.

So how were Cape fishermen employed? The answer comes with a fresh look at the nineteenth century and a new interpretation of the records. Technological advances combined to bring prosperity back to the Cape through fishing. The well-documented twist is that the fish at the center of this story is not cod, but mackerel.

Chapter 12
# Three Humble Fish

Winter had passed, spring had finally arrived, and the *Swamp Yankee* was back in the water. Captain Steve and his mate, Kerry, invited me and a few other friends to join them for fishing in Cape Cod Bay. The mackerel had returned. We picked up a few mackerel jigs so everyone could give it a try, got some lunch, and departed Rock Harbor on the high tide.

The mackerel jigs were a series of five tiny hooks on short lines called gangions attached along eighteen inches of monofilament line, and the apparatus was tied to the end of the line on each fishing rod. Each unbaited hook was festooned with an offset, elongated diamond-shaped piece of shiny plastic film and a group of short pieces of shiny thread-like material that acted like a plume.

We headed to Billingsgate Shoal where Steve put the boat in neutral and we drifted. It wasn't slack water, so we added lead flounder weights to counteract the tide and to get the jigs vertical in the water, the successful method developed years ago to catch mackerel.

It wasn't long before we got a hit. We hauled up the jig and found it had two mackerel out of five hooks. As soon as one rig was hauled up, another had fish. With a rush of adrenalin, the blitz began – put jigs in the water, haul them up, take the fish off, and throw the jigs back in the water as quickly as possible. It was all happening so fast that all we could do was laugh and banter during the forty-five minutes of raucous action.

≈≋≈

Mackerel are no longer considered as important a fish on the Cape as they once were. In the nineteenth century, hundreds of boats fishing for mackerel, not cod, filled Cape harbors. Mackerel, along with herring and menhaden, were all caught close to the Cape's small harbors, not far offshore. These three humble fish provided the necessary economic boon for Cape people to survive.

Mackerel had long been a significant but challenging species to fish for. They dive deep, swim very fast below the surface, and are rarely seen. Hook fishing was inefficient. The small hooks of mackerel jigs were lightweight, and the lines tangled frequently. Before the age of electronic fish-finding machines, simply locating the mackerel was difficult. Also, despite their vast schools, their abundance fluctuated widely from year to year. "In good years the fish may appear in almost unbelievable schools, miles in length," while in other years, "the fishery will be a flat failure," according to Henry B. Bigelow and William C. Shroeder in *Fishes of the Gulf of Maine*.

**Mackerel Fleet, Wellfleet. (Courtesy of Cape Cod National Seashore)**

Improvements in gear and mackerel fishing techniques meant a huge expansion in landings during the early nineteenth century. Cape fishermen improved their methods by adding lead to the hook shank so that the weighted hooks sank, reaching the fish below the surface, and the lines did not foul as easily. Mackerel were then caught in enormous quantities. Landings, especially from boats out of Wellfleet and Truro, filled a growing demand.

Baiting up Trawl, Provincetown, Mass.    Pub. by The Provincetown, Advocate

**Baiting long lines on shore. Gear is also known as tub trawl.
(Courtesy of Cape Cod National Seashore)**

The increase in mackerel landings coincided with great offshore cod fishery improvements in efficiency as well. Instead of crewmen standing individually at a ship's rails, each with a baited hook and line, they launched small double-ended boats known as dories from the schooners. The men in the dories scattered away from the schooners, increasing the territory the crews covered. Cod fishermen progressed to tub trawling, also known as longlining, from the dories, vastly multiplying the area of bottom covered by any one schooner. Longline fishing – miles of baited hooks set on the bottom – attracted tremendous amounts of cod even farther away from the schooner. Newer vessel designs and improved anchoring devices also enabled the large schooners to fish more safely in the notoriously dangerous but unbelievably productive waters of Georges Bank.

Yet, with all the improvements to cod fishing, cod landings on the Cape were not huge. In *Clearing the Coastline: The Nineteenth Century Ecological and Cultural Transformation of Cape Cod*, author Matthew McKenzie presents a masterful explanation of the changes that took place on the Cape at that time and the role of the fisheries. McKenzie examined harvest records to explain what was really happening on the Cape. It was all about the mackerel that surpassed cod in the nineteenth and early twentieth centuries. Wellfleet boasted more than one hundred mackerel boats, with the third highest landings behind Gloucester and Boston. Dennis, Harwich, and Provincetown followed suit, adding substantially to the Cape harvest. By 1916 and 1917, Cape boats that fished Georges Bank, South Channel, Nantucket Shoals, Chatham, Race Point, and Stellwagen Bank landed ten to twelve million pounds of mackerel per year, according to multiple reports.

Given the historic focus on cod, the mackerel harvest statistics do not make much sense at first. They raise the question of who would be eating that much mackerel.

Generally, mackerel, like herring, is tasty when fresh but less so when pickled, mainly due to its high oil content. But new preservation methods evolved. Pickling was replaced by icing which improved palatability and created a highly popular fresh product. By the mid-nineteenth century, trains connected the Cape to Boston and New York, with some rail lines extended to the docks, and fresh mackerel became a cheap food source in those prime urban markets.

Mackerel's chief role, though, was food for cod, not humans. It was caught for bait that allowed the cod fishery to flourish. Longlining the offshore banks required many tons of bait. What Cape Cod lacked in deep-water ports, it made up for in its inshore fisheries' ability to supply the bait to the offshore fleets. As the broad-based fishing industry grew, so did the Cape's population.

The inshore fishery produced more than mackerel. Herring and its cousin menhaden also boosted the economy.

For Wampanoags and early settlers, herring served as fertilizer, food, and possibly bait. Just as they had for cod, forward-thinking leaders in the late seventeenth-century Plymouth General Court exercised a type of resource management by recognizing the inherent value of the resources within its jurisdiction and regulated herring harvesting during spawning seasons. Individual communities continued that stewardship and zealously guarded herring in the eighteenth century, establishing their own criteria about how much could be harvested, by what method, and by whom. Officials favored their residents' subsistence requirements over export sales for the publicly-held herring runs, although the value of the fish outside the community was increasing. If the towns maintained tight control over the runs, the fish would always return the next spring, the people would be assured of fertilizer for the notably poor soil and food for their tables, and the communities would enjoy long-term survivability, wrote McKenzie.

That all changed at the end of the eighteenth century when towns began to sell their herring as a source of revenue. While most herring runs were town-owned in the nineteenth century and management prevailed, private landowners increasingly manipulated streams to create private runs as an investment, McKenzie explained. Selling the fish from the runs, before they spawned, either through private property owners or from the towns, made herring an economic windfall rather than food for the residents or fertilizer to grow more food. But it decreased the number of spawning fish.

Menhaden, known by several names in New England, including pogy, or poggie, or bunker, was the third species in the Cape's substantial nineteenth-century inshore fishery. Menhaden never became popular as food for settlers but were harvested as bait and fertilizer and also for their oil. An oily fish like mackerel, menhaden oil was highly useful in manufacturing leather, rope, soap, paint and as a lubricant for machinery. It was so versatile that it even became a viable, less costly substitute for precious whale oil. In 1874, Smithsonian fishery scientist George Brown Goode noted in his *Short Biography of Menhaden* that the "200,000 gallons of menhaden oil [produced in a year] nearly equaled that of all the whale, seal and cod oil made in America."

Like herring, menhaden is a filter feeder, an ecologically-essential member in the food web. Low on the food web hierarchy, it feeds on plankton and is itself a food source, eaten by every fish higher up the chain. Its ecological role and abundance means that menhaden is considered among the most commercially important fish of the U.S. Atlantic coast, according to several sources.

Little thought was given to managing the fish at sea or in the inshore areas with the exception of the care exercised once herring reached the runs. The lack of management beyond the runs allowed an extremely efficient fishing method, pound nets, to flourish.

The technique was different from hook and line, but it was not new. Pound nets, or weirs, had been used by Native Americans long before Europeans arrived, although the colonists did not use them until it became profitable to do so. These ingenious fish traps take full advantage of fish habits, especially those of herring and menhaden. Using a method that has not changed much with time, weirs can be made in many different configurations. Nonetheless, the basic pattern and manner of trapping fish is the same.

**Old fish weir on the flats, serviced by fishermen with horse and wagon. (Courtesy of Cape Cod National Seashore)**

In the early spring, fishermen set a long series of poles, called the "tail" or leader line, in relatively shallow water. They set them perpendicular to the shore from shallow to deep, making sure that the poles are taller than the high tide. They connect the poles with netting that blocks the fish from swimming along the shore. At the deep offshore end of the leader line, the fishermen set another series of poles and netting in a nearly circular, heart-shaped, or double balloon-shaped, "double bowl," arrangement. Whatever configuration is used is based on personal preference, according to tradition or performance. The offshore portion is fitted with netting draped across the bottom to form a "floor." If the structure has a double-bowl configuration, only the offshore portion contains the "floor" net.

*Fish Weir, Provincetown, Mass.*

**Old fish weir showing whole design. (Courtesy of Cape Cod National Seashore)**

When fish swim parallel to the shore, they bump into the "tail" netting. Unable to continue in the same direction, and wary of shallower water, they swim to deeper water to avoid the obstruction. When they reach the end of the leader line, they are within the first bowl. They always swim forward, and if they follow the net, they cannot find a way out and are trapped within. If there is a double bowl, the fish follow the same pattern and find an exit that actually leads to the second bowl, where they become trapped.

Fishermen tend the weirs daily. They first close the "door" to the bowl, then untie the net from the poles one by one from a small skiff, while other fishermen in a larger boat grab the net and haul it hand over hand up to and over the rail. The net on the bottom gets smaller and heavier as the men "purse" the fish to one section of the "floor." Then they lean over the rail and stretch as far as they can, using long-handled dip nets to grab flopping, squirming fish and empty the contents of the weir into their boat. The backbreaking work is messy, especially if squid are caught in the net. The squid squirt gooey black ink all over the men and boat.

Catching anything swimming along the shallow waters, these traps do not target one particular species. That includes herring, mackerel, and menhaden; but also butterfish, scup, black sea bass, and many other commercial species. The catch can be impressive. Noncommercial fish such as the ugly sea robins are often trapped, too, as are unusual visitors such as sea turtles. When they were more common, huge armor-plated sturgeon were occasionally caught in such nets. Lately, seals have had a field day within the weirs – though unlike the fish, they have no trouble finding their way out again after eating their fill.

**Modern weir with fishermen Kurt Martin untying net from poles.
(Photo by author)**

The fish traps eventually squeezed out the hook-and-line fishermen. The predictable habits of herring, swimming back home each year, were particularly conducive to the efficiency of the weirs. Similarly, vast schools of menhaden literally followed the leader into the waiting traps. The Barnstable County take of alewives peaked in 1882, with 837,000 fish caught. Just four years later, the catch declined to 288,000, a sure indication that fewer fish were coming inshore. The long history of fishery over-exploitation repeated itself, and the downward trend followed unabated, McKenzie pointed out.

**Tending the weir net, pulling the heavy net up to the rail, and pursing the fish into a smaller and smaller area. (Photo by author)**

The changes from single hook and line, (fishermen handling individual fish), to longlining offshore, (handling hundreds of fish), and weir fishing inshore, (many species of fish all mixed together), had profound consequences for the three fish that had meant so much to Cape fishermen. They were no longer concerned with individual fish and lost the intimate relationship to the fish they caught. Fish, regardless of species, were now a commodity. Fishing became a numbers game of pounds of fish, not numbers of individual fish. Size, spawning times, or conservative gear no longer mattered. The economics of fishing had changed the rules dramatically, McKenzie stated. Those impacts were a portent for struggles that continue in the twenty-first century.

Instead of many fishermen each doing their own thing, catching and selling locally, a few entrepreneurs controlled the volume of fish caught in Cape waters and sold the fish off-Cape. Ice and rail service provided access to the lucrative fresh-catch markets of high-volume sales. Towns still had control, but the sale of herring changed the game as the door was opened to nonresidents to cash in on the bait business. Additionally, thousands of pounds of fish, including the herring, menhaden, and mackerel caught in the weirs, made the Cape attractive to offshore-banks fishermen. Vessels from Boston, Gloucester, Long Island, and the mid-Atlantic states heading to the offshore fishing grounds stopped at the Cape, the nearest point to those grounds, and bought the bait they needed for the longline cod fisheries. The small-boat owners or operators who had been fishing out of the creeks, salt ponds, and small harbors that served as home ports – the people who made living on the Cape possible by serving their own needs – were simply left out of the equation, McKenzie added.

Although longlining persevered, alternative methods of fishing – gill netting and heavy-gear trawling – evolved. As they became prominent, weir fishing became appreciated as one of the least destructive of the three methods that replaced hook and line. Gill nets kill anything swimming that gets caught in the mesh, and if the nets are not tended regularly, they produce inferior quality fish that may have been dead for a while. Heavy-gear trawling can seriously alter bottom habitats. The jam-packed nets also catch non-target fish, which may die although many fish survive if returned to the water quickly. Weirs, however, catch live products that can be culled on site. The cautious weir fishermen keep the target species and throw unwanted live fish, including those that are too small, back into the water.

Nowadays, however, weirs are nearly a thing of the past, a quaint reminder of a bygone era. New fishermen who desire to enter the pound-net fishery must purchase one of the rare permits that are still available. The few fishermen still tending weirs are finding that with lower fish harvests, nets seriously fouled with heavy marine growth, the impact on near-shore areas by activities on land, and no letup in the amount of required work, the effort is just too difficult to justify.

Mackerel stocks crashed in the 1970s, and today, U.S. market demand for mackerel is low. Its strong flavor is not favored locally, although frozen mackerel is exported overseas, where the fish is still popular. The major ports for mackerel landings are Gloucester, New Bedford, and Fall River. Cape Cod landings are insignificant.

Demand is still high for menhaden as bait and fertilizer. It is also converted to fish meal, a high-protein ingredient for poultry and other animal feed; its oil is now being used as a source for heart-healthy fish oil rather than for lubrications and manufacturing. Its undiminished economic importance has led to its decline from overfishing. Menhaden are now under a quota system. Plus, its ecological significance as a plankton eater remains an important consideration for management.

As for the river herring, its story continues to hold community fascination. Its long-term viability is questionable and is reminiscent of the European exploitation of sea herring centuries ago. But there is hope now, through collaborative coastal-wide management and from an unforeseen source – citizen volunteers.

Chapter 13
# Herring: Instinctual Migration

The gulls announce the arrival of herring. They circle overhead or rest on exposed rocks along the sides of the stream at Stony Brook Valley in Brewster and squawk a cacophony of protest if another gull enters their territory. It is not a deep ravine. The hills on either side are only about fifty feet high, but that is high for this part of Cape Cod.

Year after year, decade after decade, century after century for millennia, the river herring – two closely-related species, alewives and bluebacks – have returned to this stream, one of the state's most productive and scenic herring runs, to swim upstream, spawn, and then head back down to the sea. They start their journey in Cape Cod Bay sometime around the beginning of April when the water temperature is around fifty degrees Fahrenheit, and enter the mouth of Paine's Creek. The fish swim through the extensive salt marsh to a culvert under Route 6A. There the creek emerges from the culvert with a different name, Stony Brook, and the salt marsh morphs into a brackish marsh and then to abandoned cranberry bogs. Farther along, the creek meets a gentle rise in elevation. There the fish begin their ascent to the freshwater ponds they are seeking, traveling a mile and a half to their destination where they will spawn.

Rock outcroppings squeeze the water into a high-velocity gush where the herring bound through. Relatively quiescent pools along the way are respites where the fish rest until ready to resume the journey. The strong or just plain lucky ones survive. Others feed the raucous gulls that pluck the fish as they leap or while they swim more slowly in the resting pools. In *The Run*, noted naturalist writer John Hay explores the interrelationships of the fish to a world in which their role is tied to many other creatures, including us. He magnificently captures the essence of herring as they journey up the interconnected sequential environments from the sea to fresh-water ponds:

"The fish kept moving up. I watched them swinging back and forth with the current, great-eyed, probing, weaving, their dorsal fins cutting the surface, their ventral fins fanning, their tails flipping and sculling. In the thick inter-balanced crowd there would suddenly be a scattered dashing, coming as quickly as cat's paws flicking the summer seas. They may have moved by "reflex" rather than conscious thought, but what marvelous professionals they were in that!"

Some runs, such as the one at Stony Brook, evoke a timeless natural setting. Visitors meander along footpaths winding around a creek filled with water tumbling over rocks in a series of stepwise cascading falls, a natural ladder. To further enhance the experience, a still-operational nineteenth century paddlewheel gristmill perpetuates the art of grinding corn and brings one's focus back in time. The mill utilizes the power of flowing water, but the water drops precipitously from a height much too great for fish to negotiate. So a concrete structure – a fish ladder – diverts the herring and provides manmade "steps" up to the pond.

The abundance and importance of herring has long been acknowledged. Captain Charles Whitbourne, a seventeenth-century settler, reported the bounty of herring and the vast herring runs in his book *The Travels of Captain John Smith*: "Experience hath taught them at New Plymouth that in April there is a fish much like a herring that comes up into the small brooks to spawn, and when the water is not knee deep they will presse up through your hands, yea, thow you beat at them with cudgels and in such abundance as is incredible."

In 1953, Henry C. Bigelow and William S. Shroeder elaborated on herring habitat in *Fishes of the Gulf of Maine*: "When the white man crossed the Atlantic probably there was no stream from Cape Sable to Cape Cod but saw its annual run of alewives unless they were barred by impassable falls near the mouth. Alewives are decidedly general in their choice of streams running indifferently up rivers as large as the St. John, Merrimac and Potomac or streams so small that one can almost leap across, and only a few inches deep."

When European settlers arrived on the Cape, the Wampanoags taught them how to wrest the most from the poor soil by adding herring as fertilizer. Beans, squash, and corn, staple crops for the Native Americans, were what today's organic farmers would call "companion plants." The Wampanoags buried a herring, mounded the soil on top of it, and planted the seeds of all three vegetables in the small hills. As the fish decayed, it added nutrients to the soil. The corn grew tall, the strong stalks a perfect trellis for the beans to climb; the beans added nutrients, especially nitrogen, to replenish the soil; the squash meandered on the ground, the broad leaves shading the ground to prevent weeds from gaining the upper hand.

The Indians taught the settlers when to expect the herring to return and how to set traps in the creeks as the fish migrated upstream. The settlers learned the technique and farming method, and while not adept at fishing themselves, they knew herring as a food that had been important back in England.

River herring, both alewives and bluebacks, spawn in quiet water and slow-moving streams but never in rushing water. The Cape's glacial geology formed perfect habitat for them: numerous streams leading from the sea to freshwater ponds. The true number of these connections is unknown, but historic topographic quadrangle maps from any Cape town provide tantalizing clues.

After the herring are born in freshwater ponds, they swim to saltwater as very small fry and stay at sea until they reach sexual maturity at age three or four. Then they swim back to the estuaries, where the salt and fresh waters meet. In the estuary, they find their natal stream, the same one in which they were born, and fight their way up to the fresh water and spawn.

The understanding that herring return to their home creeks originated from efforts to reintroduce them to rivers that had become nonproductive. Adult spawners were released, and about four years later the next generation of spawning herring returned where there had been none in the intervening years, explained Bigelow and Schroeder.

Unlike salmon that die shortly after spawning, the spent herring do not linger in the freshwater ponds but turn around and swiftly swim back downstream, while others simultaneously struggle upstream for their turn to spawn. They require little time to acclimate, able to regulate the osmotic difference in their tissues between salt and fresh waters and then back to salt with incredible facility. Herring may return to their home creeks a second time a year later, but scale analyses show that only a small percentage live to spawn multiple times, John Hay writes.

At the Stony Brook run, the herring that survive the journey to spawn swim back down the stream to the mouth of Paine's Creek. If they get there at an ebbing tide, the creek fades away at the mile-wide exposed flats, leaving the herring as fair game for gulls flying overhead. After the effort of getting up the creek to spawn, they come to this fate. As the tide floods, the fish that come after them have enough water to get out to the deep bay and offshore beyond.

**Stony Brook Herring Run in Brewster with gulls lurking on the shore for a meal. (Photo by author)**

Back in the home ponds, the herring eggs incubate for about six days, with water temperature determining when the eggs hatch. A month later, when they are about three-quarters of an inch long, the tiny fry scuttle downstream to the estuaries where they stay until they are at least three inches. Somehow, during their trek downstream, they imprint on their surroundings for future reference so they can return as parents. But summer is a time to eat or be eaten. Many of the tiny herring become food

141

for an immense number of species within the estuaries, and then they must make it through the ordeal that follows at sea. If successful, then they too swim offshore, explained Bigelow and Schroeder.

On a late April morning in the late 1990s, at another run a few miles from Stony Brook in the Pleasant Bay estuary, a small, shallow pool rippled with fish milling around, swimming lazily, not frantically. The fish had come in to the salt pond. Gulls flew over the pond and congregated on the water's surface. The herring had made it to the pool at the base of this run on the nighttime high tide, and early in the morning they appeared to be resting, clearly visible from a bridge that crosses the stream.

Suddenly, one fish darted through a culvert under the bridge, and then another, as if some mysterious cue had led them to get in line to make their ascent. The herring began to swim up the concrete ladder on the other side of the culvert while gulls tried to abruptly end their destiny. The ladder, not nearly as pastoral as the strategic rocks at Stony Brook, is intended to help the fish swim up the fifteen-foot rise in the creek bed. At each "rung," it creates a mini-waterfall and strong current to the next slightly more elevated rung. Every enticing waterfall empties into a small pool where the fish rest during their journey upstream before negotiating the swift current of the next waterfall.

**Concrete ladder at Pleasant Bay herring run. (Photo by author)**

At a nearly level section of the creek, the fish swam and wriggled over the shallows where barely an inch of water flowed over the sandy bottom. Then they continued to the twelve-acre freshwater lake. There, a concrete "holding pen" with removable slatted "pen boards" regulates the flow of water coming down from the lake. Once through the turbulent water within the pen to the more quiescent water on the lake side, they reached their goal to begin the cycle of life.

After they spawned, the gushing water of the fish ladder, where the fish had expended so much energy swimming against the flow, was in their favor as they raced back to the salt pond at the base of the run. Not quite as salty as the ocean, the pond afforded the herring a chance to adjust from freshwater back to salt. They did not stay there long, but found their way along a narrow river that meanders through Pleasant Bay to Chatham Inlet, nearly fifteen miles from the salt pond. After passing through the inlet, they disappeared to join the giant schools of herring at sea.

Mills like the one at Stony Brook in Brewster utilized flowing water in the early days. However, they were required to open sluiceways to allow free passage for the herring or they were prohibited from operating when fish were heading upstream or down during spawning season, explained John Cumbler. If there wasn't a way for the fish to avoid them, the herring could be injured or die at the mills.

By the early nineteenth century, Cape towns faced "conflicts between those supporting common-law rights to the fish and those who saw the future in laissez-faire economic expansion," according to Cumbler in *Cape Cod: An Environmental History of a Fragile Ecosystem*. That economic expansion included building dams to create power for mills or making flood-prone areas suitable for building. The reactions to these barriers exhibited a classic polarity between those arguing for strong private property rights and those who supported common resources available to all. It was also a conflict between groups of people with different economic viewpoints. The advocates for dams were perceived as encouraging economic prosperity. Those advocating for the natural resources, the herring, were perceived as supporting poorer citizens who needed and utilized the fish. That type of social dichotomy — economic development versus environmental conservation — remains, though not necessarily overtly.

143

As the Cape developed, other obstructions appeared in the form of roads and land development projects. A blocked passageway that removes one pond as an active spawning area is, by itself, probably inconsequential. But multiply such obstacles on all the creeks, streams, and rivers on the Cape and throughout the spawning range of the herring, preventing them from reaching their instinctual spawning grounds, and the habitat loss is profound for the population as a whole.

While herring have been depleted from overfishing, habitat loss has been equally destructive to the population as the number of functioning herring runs has dwindled over the years. Filling, damming, rerouting, and otherwise altering creeks for commercial and residential development has destroyed native habitat and prevented herring and other fish from performing their ancient rituals to reproduce. It is nearly impossible to successfully replace that habitat after the fact, unless redevelopment, which includes restoring altered environments to their original conditions, is part of the equation. Thus, the herring story is another aspect of the tragedy of the commons. Yet, groups continue efforts to reclaim habitat by restoring herring runs such as the Shawsheen River in Andover, Massachusetts, where, in 2017, the first herring were returned after a two hundred-year absence when the last dam was removed. And one of the largest restoration projects in the Northeast, the Herring River in Wellfleet, is heading for the permitting phase after exhaustive planning.

At present, river herring are dwindling fast and are listed as a federal species of concern. Since 2005, the Massachusetts Division of Marine Fisheries, the agency that monitors river herring populations, has banned catching, selling, or possessing them. The ban is part of a management plan for river herring by the Atlantic States Marine Fisheries Commission.

The commission was an early regional fisheries management group formed in 1942 and chartered by Congress in 1950 in recognition that "fish do not adhere to political boundaries." It serves as a deliberative body coordinating conservation and management of the states' shared nearshore fishery resources – marine, shell, and anadromous – for sustainable use. The commission develops plans for twenty-three species. It spearheaded a controversial but successful Atlantic Striped Bass Conservation Act of 1981 that heralded the recovery of striped bass populations beginning with size limits and moving toward a multistate moratorium. Building on that success, the commission entered into a partnership in 1993 with the National Marine Fisheries Service and the U.S. Fish and Wildlife Service resulting in the Atlantic Coastal Fisheries Cooperative Management Act. It required that all states that are included in a fishery management plan developed by the commission must implement required conservation provisions. Noncompliance may result in a federally-mandated moratorium on fishing activity in that state.

Thus, the Massachusetts herring ban is part of a larger management plan under the marine fisheries commission where a current moratorium on harvesting is one of the requirements. Unfortunately, offshore pair trawling for sea herring, a related species, continues to take river herring, a problem that is being addressed.

Towns bear the responsibility for maintaining their runs and take that responsibility seriously. Nearly every town has a Herring River or Herring Pond or a Herringbrook Road, another clear indication of the importance that herring runs have had in Cape culture. Town "herring wardens" are no longer alone in their obligation to maintain the runs, though. The public has embraced the idea of lending a hand.

A small army of volunteer citizen scientists monitor the population each spring, acting as official herring counters at these long-used Cape herring runs. The volunteers, clad in warm jackets, hats, gloves, and waders, and armed with rakes or shellfish scratchers or other appropriate gear, tromp through the runs in March, before the fish arrive. They remove branches, leaves, detritus, trash, and anything else that may have accumulated in the run that would impede the fish that are coming – or that they hope are coming.

They take pride in what they do. They cannot use the fish for food or bait since taking them is prohibited for the time being. They know, though, that the herring could be a meal for a striped bass or bluefish heading that way, or may also feed the cod, haddock, or other types of commercial fish.

The folks who volunteer feel that giving the herring a chance to keep the population healthy is the right thing to do.

In a good year, counting fish should be an exacting task that takes substantial concentration, with volunteers working throughout the day in short shifts of no more than ten minutes to count the many individual darting, squirming fish, in order to record and document the extent of the run. Since 2005, when the moratorium was imposed, it has been anything but a difficult task.

At the Pleasant Bay run, the volunteers stand on a little wooden bridge over the narrow run and count the number of herring that swim below. There are usually few fish to count, and many times, none. Even when there are no fish, volunteers take their turn at their appointed times. Standing at the herring run and counting zero herring week after week is not much fun for the counters, but the data will be incredibly valuable if or when the herring rebound. It will document and identify the start of a change, critical information that will later be analyzed to determine the health of the population as a whole and to project the future of the species.

Now, the zero counts speak to the tragedy that has struck the inshore fisheries and the larger concerns of the North Atlantic fisheries. Yet the volunteers continue their task. What drives them to go to the run with low expectations? For some, it is the tranquility of the run itself. As volunteer Jane Hussey wrote to me: "There is no pool on either side of the little bridge that we stand on – just a serene, shallow four to five-foot-wide stream that runs four to six inches deep. It is shaded by trees and has ferns up and down, wild or invasive roses, etc., sandy bottom that we muck out each March, so the fish have a clear passage from leaves and sticks. A little brass or bronze thermometer is hanging down from the wooden bridge via nail and string so we can measure the water temperature. There is another thermometer mounted on a garden stake, not facing the sun, so we can record the air temperature."

For others, it is a lifetime of observation. Another volunteer, Pam Herrick, grew up on the pond. "My earliest memories of the herring in our run were in the early 1940s. As a child, my memories and imagination mixed together, and the herring running with the gulls sweeping remains as a wild and exciting memory. Fishermen came with buckets, which they scooped into the stream and filled with herring. These herring were used for bait. My family tried to eat herring, but the bones were too big a challenge. Much later in my life, I discovered the women in Provincetown had delicious recipes for herring, and probably still have these. Someone should collect them, as they may turn into historical documents. Also, as an adult, I became concerned with the bucket harvesting, and not long after that the ban on taking herring took effect. In those early years herring were just a part of spring, and we took them for granted because we knew so little about them."

For still others, it is the hope of one day going there and finding fish swimming up the run. Kathy Whitelaw expressed it this way: "So sorry about over counting. They just kept coming in swirls. It was a ballet!!!! In my three years as a counter this was the best, a Broadway production!!! In truth, it is a ZEN moment at that little bridge in late afternoon sunshine!!!"

On June 22, 2015, Judy Scanlon, the herring volunteer organizer, coordinator, and data collector for the run, explained: "The young of our count subjects are now starting to leave the lake and migrate to the estuary. You can see them in the flume section between the two pen boards at the top of the run. We also observed some yesterday at the bottom of the run. Another person reported that they were at the top of the run last week. I recall that last year the water level was so low in the lake and run that many could not leave the lake until November." The earlier exit in summer is more typical, and a hopeful sign for all concerned.

※

Even on the zero-count days, the run is not empty. A closer look reveals another of nature's mysteries. Tiny squiggly forms mill around. They are baby eels or elvers, also called "glass eels." According to Jane: "Someone ingeniously put white 12 x 12 clay tiles on the bottom to make it easier to see the fish and elvers that go by."

The eels are catadromous fish, the opposite of the herring's anadromous lifecycle. These mini eels are spawned not in the freshwater lake, but rather in a large oceanic area thousands of miles away, the

Sargasso Sea. The baby elvers swam to this creek from the middle of the North Atlantic off Bermuda, though how they choose a particular stream is unknown. The eels live in the freshwater and estuaries until they are mature, when they will return to the Sargasso for their own procreation rituals.

Declining by over 90 percent in the last two decades, eels are also in much shorter supply than in the past. Unfortunately, they are especially valuable as glass eels, worth a tremendous sum as a popular delicacy in Asian markets. It is legal to catch them based on a quota system in Maine and North Carolina, but it's illegal to take them in all other states under the ASMFC American eel plan. So their value creates a huge poaching problem. In March 2017, seven individuals were indicted for violating federal laws against trafficking in illegal elvers in Massachusetts, North Carolina, New Jersey, Delaware, and Virginia.

Pam reminisced about the eels of her youth, too: "We also caught eels in traps tied to our dock. There were lots of eels, also, in those early years. We skinned them and fried them up in a frying pan and most of us as children actually liked them, even though we made gagging noises as they were put on our plates."

Eels are also counted at some of the spawning streams because of their own decline and the pressures on them. Seeing the tiny eels in the creek is a hopeful sign for the future – if they can make it up that last step on the ladder to the comparative safety of the lake without being taken.

~~~

Herring are at the center of a swirling controversy. Although not widely understood and appreciated by the general public for their role, they are recognized lynchpins in the food web. They are a potential meal for every predator in the sea – striped bass, bluefish, cod, trout, sea birds, osprey, blue heron, seals, and otter, to name some. With so many predators, the herring's survival rate is poor, and only an estimated one percent survive the journey back to sea. In the seas, those that make it to adulthood swim in schools of fantastic numbers. For protection from natural predators, there is safety in numbers. But when man is the predator, the fish's schooling behavior only makes it easier for large pair trawl nets to trap enormous quantities.

Dams, improperly constructed or poorly maintained culverts, clogged waterways, and pollution of many sorts have had enormous consequences, leading to declining numbers of fish to reproduce. This habitat loss, along with overfishing with weirs at the runs and with giant nets at sea, are all reasons that herring are in trouble. Finally, as we begin to understand some of the intricacies and interrelationships in the marine world, the decline and the causes become obvious.

In a petition to list alewife and blueback herring as threatened species, the National Resources Defense Council listed examples of species declines in major rivers along the Atlantic Coast: a collapse of herring in the St. Croix river on the border between Canada and the United States; a 90-plus percent decline in the Connecticut River, the Chesapeake Bay, and North Carolina's Albemarle Sound; an 85 to 95-percent decline in Massachusetts' two largest remaining runs; and total absence in South Carolina.

Atlantic, or sea, herring are a different species, closely related to the river herring – alewives, bluebacks, and menhaden – but do not require fresh water for any part of their lifecycle. The formerly abundant river herring have declined drastically. At sea, the species often overlap in their patterns, and now, with the precipitous decline in river herring, the sea herring that were formerly not as highly exploited are harvested heavily.

In medieval times, European herring populations crashed, possibly from overfishing or a combination of fishing pressure and climate change that resulted in the fish not being where they had always been located. If overfishing was the primary cause, it is nearly incomprehensible to us now, given the limited technology of the time. It seems we never learned the lesson, because it is likely that the current herring populations are or have been in jeopardy because of us.

When the herring are in trouble, everything above them in the food chain is also affected. They are needed to maintain healthy populations. Since herring are at the base of the food chain, their loss means cascading effects through the entire food web, with huge implications for other fisheries.

Several decades ago, we did not think much about the effect that individual fishing efforts had on large populations. No one paid much attention to the number of fish taken from any particular stream because more would come to spawn the next year. Fishermen from local harbors were only interested in what they brought home, not what other fishermen from other ports did, as long as there were fish when they went out again. Once offshore, everyone competed for as much fish as they could safely haul in their boats. What difference did it make if one stream was blocked and the herring could not return to spawn? Few people considered the larger picture in relation to population dynamics.

Fishing was very different then. That is not the case now.

# Chapter 14
# Growth of an Industry to Devastation

Soon after midnight on a cold April night in the 1960s, the boat left Chatham Harbor, as it had done for years. As always, the captain approached the dreaded Chatham Bar with great caution. He slowed the boat and listened.

The opening to Chatham Inlet is a nasty piece of water. It is wider than that of Nauset, eight miles north, but just as dangerous. The captain feared the bar as he always did – fear mixed with respect for the power of the sea – but that dread could not govern his emotions. If it did, he would never make it as a fisherman who had to negotiate the inlet twice daily. Above the drone of the engine, the captain could hear the surf, endless lines of waves breaking on the sometimes submerged sandy bars. The vicious surf calmed at the narrow channel through the sand bars – a channel that moved daily with the swift currents.

The captain was confident he could find his way through the channel in the pitch black. Breakers tended to occur in sets of three, with the last one often the highest wave. He knew there would be a lull before the next set. And he knew he had to time it right to pass through the inlet safely.

In the darkness, the captain called on all of his senses. His acute hearing picked up nuances in the sound of the waves that crashed all around him and on the distant beach. He felt the rhythm of the surf. The crested wave's foamy surge lifted the boat up and over. He felt the power of each wave as the energy transferred to his feet planted firmly on the deck. While every vessel reacts to the water's movement differently, he knew his boat and could judge what was happening.

While he listened and waited, he thought back to the days when he was a boy and he and his friend would go flounder fishing.

～～

The sky was crystal clear and sunny with a first nip of autumn in the air, although the water still held some warmth. The boys brought homemade gear: a hand line – a hook and lead weight tied to the end of a length of twine wrapped around a four-inch square wooden frame – and extra hooks just in case; plus a small bucket, a quahaug scratcher, and a clam hoe to dig clams or clam worms – all of them good for bait. Clams were harder to dig but easy to find, and the brittle shells made them easier to crack open than the hard-shelled quahaugs.

The flat-bottomed skiff rested on the silty sand. The boys picked up the anchor and shoved the boat into the water. The future captain leaped in first. He grabbed the middle seat, slamming the oars into the oarlocks while his friend dropped the fishing gear in the boat. Then his friend pushed the bobbing boat farther out into the water before he jumped into the stern.

The flounder were running. The boys had been out on the bay every chance they could to watch for the blackback flounder to return on their way to the ponds, their winter homes. The fluke, or summer flounder, were already gone, heading south. But the winter flounder, which had arrived recently to spawn in the coves and salt ponds, were even tastier. Fall flounder were big. And fat. They had been feeding out in the ocean since spring. All a boy had to do was set the anchor near the channel, put a clam or piece of quahaug on the hook, tie a lead weight to hook and line, let the hook sink, and wait to feel a gentle nibbling sensation on the end of the line. Then, at just the right time, he would give his line a quick tug to set the hook and haul in a flapping flounder.

The captain smiled as he remembered that he had done better, catching more fish than his friend. His friend studied his technique, yet could not figure out what he was doing differently. Finally, the captain revealed the secret his father had taught him. "Knock three times," he said.

"What do you mean?" his friend asked.

"You feel the weight hit the bottom. Once it hits, pull it up just a bit,

152

let it hit the bottom again, and then a third time. Then hold the line so the hook is just off the bottom. Hitting the bottom a few times gets their attention, and then the flounder go after the bait. Works every time."

His friend tried the new method and wham! He got a hit almost immediately. Then another, and another.

When the fish stopped biting, the boys rowed further up the channel to another favorite spot. The tide was flooding, running harder than when they had headed out. They drifted with the tide instead of anchoring, to see how that would work. Knock, knock, knock – jerk! Another slammer. Then another. Finally, reluctantly, they hauled in their hand-lines and rowed back. It was an easy row going with the tide, and soon they were near home, where they anchored the boat and headed up the path.

A couple of days later, the boys switched places in the boat. They both liked to row. They rowed from the landing out into the shallow water around the island and toward the harbor. The tide was ebbing that afternoon. Closer to the harbor, the wind picked up. Waves tipped with white replaced ripples as the water became choppy. It was tough to row against the wind and tide, and they knew it would be hard to get around the point to make it back home. But they always carried a mast and small sail in their little flat-bottomed boat. Temporarily they returned to shore, put the mast in the hole in the bow seat, hoisted the sail, and transferred one of the oars to the U-shaped notch in the transom to serve as a rudder. Then they shoved off again. Sailing got them home safely. They lowered the sail and anchored the boat on the shore.

Young boys on Cape Cod may not have realized that their play and exploration would set their course for life. As they rowed and sailed the waters of Nauset, Pleasant Bay, or any of the numerous other Cape bays, they learned about local currents, the forces of tides and wind, water movement that affected the boat, and how to safely maneuver in that watery world. They acquired the foundation of skills they would rely on for a lifetime on the water.

Nantucket native Walter Folger described in 1791, the effect of whaling on Nantucket boys who dreamed of becoming whaleship masters, noting in *A Topographical Description of Nantucket* that when local boys are very young, they make use of common whaling phrases, and "as soon as they are some years older, they are seen rowing in boats for diversion, which makes them expert oarsmen, a thing that is requisite in taking a whale." A few years later, in the early nineteenth century, Yale president Timothy Dwight was struck by a similar sight of boys around the Provincetown shore. In his book, *Travels in New England and New York*, published posthumously in 1823, he wrote: "Little boys managed boats of considerable size with the fearlessness, and apparently with the skill, of experienced boatmen. Every employment, except within doors, seemed to be connected with the water and intended for the sea. To fish in every various manner, to secure that which had been caught, to cure fish, to extract oil, and to manage different sorts of vessels, from the canoe to the ship engrossed apparently the whole attention of the inhabitants." And Massachusetts Representative John Gorham Palfrey offered his impression of local boys and boats, stating at the 1840 bicentennial celebration of Cape Cod's settlement: "The duck does not take to the water with surer instinct than the Barnstable boy…. It is but a bound from the mother's lap to the mast-head….He can hand, reef, and steer by the time he flies a kite." Thus, the Cape and Islands produced some of the most knowledgeable and skillful mariners for fishing, whaling, and maritime trade – the seamen who manned the boats, and the ship's officers, many of whom who rose to become shipmasters.

In the decade or so following World War II, young boys growing up on Cape Cod continued the tradition of rowing and sailing in the bays, imagining adventure, exploring their surroundings, learning of the animals and plants, and discovering what ate what and what hid from what. They tested baits to see what worked best for flounder, bass, and bluefish. They fashioned lures alongside their fathers, becoming better as their fingers became nimble. They gained valuable experience and perfected fishing techniques on every small voyage out on the water. They learned which holes in the sandy mud were made by clams, by quahaugs, and by sea worms.

Authors with Cape names like Eldridge, Sparrow, and Nickerson all evoke the flavor of the times on the Lower Cape in the early to mid-twentieth century. No one of that generation could imagine that by the late twentieth century, flounder fishing would be a distant memory. So would hunting, gunk-holing in the marshes, and even messing around in a small rowboat in any of the Cape's estuaries. The Cape and its abundant marine resources were about to change beyond their wildest imaginings.

The boat captain heading out from Chatham waited patiently, alert to subtle instinctive changes that told him to go or compelled him to hover and wait longer until he felt it was safe. Finally, he sensed that the time was right and pushed the lever to full throttle, taking advantage of a lull just long enough for the boat to get through the Chatham Bar.

With his vessel and crew safe from the pounding surf, the captain set his course for the fishing grounds. They headed south toward the Stonehorse lightship, near the tip of Monomoy, the first waypoint to guide him to his destination. The lightship was one of several manned Coast Guard vessels anchored at specific locations to aid in navigating around tricky areas such as Handkerchief Shoals or Pollock Rip off Chatham.

The captain's compass, chart, watch, and sounding lead, along with his knowledge of his boat and his skills honed by a lifetime on the water, told him his position. Out of sight of land, where there were no range markers, he relied on the tools of his trade. Getting the boat and crew safely to productive fishing grounds and home again was a captain's responsibility. But he had no electronic navigational gadgetry on board. Radar, an essential component of military, Coast Guard, and commercial transport vessels by the early 1960s, was too costly for the owners of the typical forty-foot fishing vessels out of Chatham, and Loran, long-range navigation, was still out of the question, too.

Some of the fishing grounds covered only small areas of the ocean bottom featuring the right type of habitat. The sounding lead, or lead line, was essential for finding the correct location. A sounding lead is a device that measures depth and provides visual evidence of the type of bottom beneath the boat. A heavy metal object resembling an upside-down ice cream cone is tied to a long line, knotted along the length at every fathom. When lowered to the bottom, the greased convex portion of the lead, the top of the cone, picks up the impressions of the bottom – sand, pebbles, cobbles, mussels, or other indications of bottom type. Each species of fish prefers a particular bottom type, so through repeated soundings – lowering the sounding lead to the bottom with fresh grease – an expert captain could literally "read the bottom." With that visual confirmation, in addition to his chart, compass, course, the boat's speed, and measurements of how long it took to travel between markers such as lightships anchored "on station," the captain knew without doubt where he was and what he would catch.

The crew had different responsibilities. There is no rest at sea. Before reaching their destination in the early morning, they baited thousands of hooks with the two hundred pounds of frozen bait that had been loaded before they left port and neatly coiled the baited hooks and longlines into a tub. They arrived at the fishing grounds by dawn, and when the captain was satisfied about the location, he gave the order to set the gear. The crew was ready. They tied a baited longline to an anchor and threw the first flag buoy and anchor over the side. They then guided the longline over the stern, ensuring it did not get tangled as it payed out, then tied a second tub of gear to the first and guided it as well, and threw a second anchor and flag buoy overboard. That first line "soaked" on the ocean bottom to entice the cod or haddock while they set the next string. When all the tubs of baited gear were set, the men took a short break and then went back to the first set to haul the gear.

A skilled crew could catch thousands of pounds of fish less than fifty miles from homeport. Longlining or tub-trawling, as this type of fishing was called, had not changed much in centuries, since dory men first set lines of hooks a fair distance from the schooners. It was an efficient method of catching cod, haddock, halibut, and their ground-fish relatives then, and it was still an efficient method in the 1960s.

The secret to getting the lines in the water as soon as the crew reached the fishing ground lay in adequately preparing the gear during the night.

Just one set of gear is three thousand feet of line, over a half mile, coiled in a wooden tub somewhat larger than a bushel basket. The crew set between five and eight tubs on every trip. Every six feet, gangions, or smaller lines, are tied on, each with a hook on the end. This night, the crew baited four thousand hooks with two hundred pounds of bait cut into one-inch cubes during the four hours it took to travel to the fishing grounds. The crew needed to be wakeful, attentive, fast, and careful, as sure-handed quickness and precision are essential.

Every crewman appreciated the skills of the others. It was critical to know exactly where a hook would be in the coiled gear to prevent getting stuck while baiting a hook. Charles "Tiggie" Peluso fished out of Harwichport and Chatham for over three decades, from the 1940s to the '70s, and was an acknowledged expert. When Tiggie reminisced about baiting twenty hooks in a minute, he described the blisteringly fast repetitive motion used by nimble fishermen to bait a full tub of gear, five hundred hooks, in twenty to twenty-five minutes.

Fishing was not steady money. Weather, mechanical breakdowns, poor harvests, or poor prices all affected a crewman's take. It was a long winter if the harbor iced up, and the fishermen knew they had to go over the bar as often as they could to make enough to get them through the season.

It is difficult to comprehend how these captains and crews were so astute at knowing the environments where they earned their livelihoods. The fundamental and essential skills of those fishermen are nearly lost now because technology has taken over. Fishermen no longer need the years of experience; machines replace that experience-gained knowledge.

While we might mourn the loss of the old skills, the safety and efficiency of using electronic equipment to find and catch fish seem worth the price to many mariners. Still, at sea, if the unimaginable happens and today's electronic tools of the trade – radar, fish finders, depth sounders, GPS, cell phones – all fail, then those skills once so important and common a half century ago can be mighty handy.

Fishermen of Tiggie Peluso's generation had no idea that they were on the cusp of one of the most important eras in fishing, almost as important as the change from sail to steam. Their style was "seat-of-the-pants" hook-

and-line fishing, where self-reliance was the most important trait for economic and personal survival. The amount of fish hauled into a boat depended on the crew's stamina and how much the boat could hold. Hopefully, they would get a good price at the dock for their effort.

By the late 1960s, change was evident everywhere – on the land as well as on the water. Small houses became more rare as waterfront land was gobbled up for large, expensive homes, and the old Cape cottage became a relic of the past. There was a post-World War II building boom, but it did not reach the Cape Cod elbow until the early 1970s. When it did, "No Trespassing" and "Private Property" signs that dotted the landscape prevented access to the water, limiting freedom to launch a boat or hunt or fish wherever one chose.

The changes were so great that by the time postwar children matured to adults, flounder were in trouble and the Commonwealth of Massachusetts stepped up its management responsibilities, enacting flounder fishing regulations meant to protect spawning fish. Yet, flounder fishing inside Nauset and Pleasant Bay diminished noticeably. When humans are the predator, the most common result is that, eventually, the number of fish dwindles.

By the last quarter of the twentieth century, there was a profound change in the use of the waters. Most boys no longer passed time in rowboats but waited impatiently until the magic day when they were allowed to go out in motorized runabouts by themselves. They zoomed up and down the bays just for the joy of riding in a motorboat, and as they got older, they added knee boarding, tubing, or waterskiing. Some went fishing, but the thrill of driving a boat long before they could legally drive a car was more important to many more. The speeding boat was exhilarating and gave them a sense of freedom they could obtain no other way. So for most of them, driving a boat at an early age was the beginning of a lifelong love affair with boats and fishing. Recreational fishing and running boats became a rite of passage.

Recreational fishing is a multibillion-dollar industry – $800 million in Massachusetts alone, with an estimated two-thirds of that revenue coming from the Cape and Islands. Tourism is a huge boost to the economy.

Recreational fishing keeps the marine supply and tackle stores in business and raises revenue at ancillary businesses such as vacation rentals, inns, and restaurants. When sport fishing is added to commercial fishing, the impact on the fishing industry is enormous.

Something seems to have been lost with the change from wind and people power to engines and gadgets. The water became a beautiful scene to look at, a way to convey people from one place to another, a means to an end, and a wide-open place of opportunity. Only when a boat stops or drifts at a fishing area does anyone on board have the chance to pay attention to what is in and around the water. Speeding on top of the waves does not necessarily convey the same connection with the bay or the creatures that inhabit it that a slower, more participatory and methodical observational approach provides. There is joy, thrill, and exhilaration, but no chance to discover and watch the intricacies.

The bays became a beautiful vista, but the connection between the surface and what lies beneath became less and less important, as did the understanding about the pieces of this puzzle and where we all fit in.

For the commercial fishing fleet, technological advances invalidated the old style of fishing. Longlining is still used, but the gear has been improved. Synthetic lines no longer have to be so meticulously coiled, the tubs are baited on shore rather than on the boat, and a machine guides the lines overboard. Small-boat longline fishermen like those from Chatham compete with fishermen from larger ports with bigger boats and different methods of fishing – gill netting, trawling, and dragging.

Yet the changes in Chatham or even the larger ports were puny compared to what was happening in the North Atlantic in general. The introduction of ever more sophisticated electronics and other technological advances increased harvests, but they could not prevent what happened to the once-vast fish resources near U.S. shores. No one could imagine or predict that those exciting technologies that commanded global fascination would be so intimately tied to almost fatal overfishing. And no one in the early 1960s could predict that the cod stocks that helped spawn a new nation would collapse. No one could foresee that fishing would become a matter of regulations governing how much fish could be caught and where, rather than individual ability and desire.

How did this occur? It took place because of a continuation of a centuries-old attitude and ocean-wide myth: that the seas are inexhaustible and we can take whatever we want. If one species was overfished or failed, others remained to be exploited.

Technological advances during the twentieth century outpaced the fishes' ability to adapt to new threats to their existence. There was no place to hide. Technology provided the unique ability to pinpoint sources of fish, not just confirm the undersea terrain that might draw them as had information gleaned from the sounding lead. If there were fish to catch, technology allowed captains to find them faster and more surely. The industry escalated from a tough business in which hardy fishermen spent days at sea, labored intensively, and got little rest in order to feed their families and sustain a community, into something unforeseen. And it happened quickly.

As a result, fishing as a way to make a living – handed down through generations – was no longer the desirable career choice it once was. Clashes with biological realities, economic incentives and disincentives, regulatory "fixes," social responsibilities and attitudes, all swirled in undercurrents around the fishing industry that continue today.

## Chapter 15
# Invaded Territory and Clashing Schemes

Trips to Georges Bank were infrequent for Chatham fishermen in the early 1960s. More than a hundred miles was a long way to go with their forty-foot wooden boats. But if fate smiled on them and the weather held, perhaps they would load up with prized halibut and get back while the price was still good. The fishermen needed to trust themselves and their boats to steam twelve hours to get to the grounds. They loaded a ton of bait – literally – and food enough for three days. A boat with a gasoline engine couldn't hold enough gas in its tanks. They had to carry more in fifty-five-gallon drums, executing the tricky maneuver of filling the tanks on a rolling sea. The sporadic trips to Georges also meant less knowledge about the ocean bottom, so repeated soundings were essential to find the right kind of habitat of black and brown pebbles with bits of coral growth on them. If the fishermen were lucky, they might find enough big fat halibut to make the trip worthwhile.

As the 1960s progressed, Cape fishermen from Provincetown to Chatham to Falmouth, and Massachusetts fisherman from Gloucester to New Bedford, began to see a troubling sight – cities of lights at night, far from land on the open ocean. They turned out to be hulking, enormous ships from foreign ports. These ships roamed the North Atlantic, fishing in what had always been considered U.S. waters. The days of wooden boats and sounding leads, big fat halibut, and catching as much fish as the boat could hold were coming to an end. But the fishermen didn't know it then.

Precise determination of the boundaries of a nation's waters had been debated for centuries. Coastal countries considered the territorial sea as much a part of the state as the land itself. Traditional territorial jurisdiction extended three nautical miles from shore, the distance that could be reached by a cannonball fired from land. Later, international law gave countries the legal impetus to establish their seaward boundaries. In 1799, some countries used that legal authority to increase the boundary to twelve miles and justify boarding vessels. By the mid-twentieth century, the U.S. Congress enacted legislation allowing federal and state control of the continental shelf beyond three miles where oil, gas, and minerals were important considerations. Fishing interests were not included, but exploitation by foreign nations of fisheries close to the U.S. was troubling.

Boundaries were problematic elsewhere. Seaward boundaries and sovereign rights were resolved by defining the territorial sea, contiguous zone, and continental shelf through international treaties in 1958. But not fisheries zones. It took nearly another thirty years before the definition of an exclusive economic zone (EEZ) that included fisheries became effective. In 1982, it was codified in a document known as the United Nations Convention on the Law of the Sea. But the U.S. didn't wait for the codification. It established a two hundred-nautical mile EEZ in 1976 after more than a decade of witnessing the extraordinary efficiency of factory ships so close to its shores.

England was the first country to design and build a factory trawler, a vessel with the capacity and equipment to catch and process fish at sea. She was called the *Fairtry*, built in 1954. With crew comforts unusual for fishing vessels and all business equipment below decks, she was designed to remain away from port for months, to land and process not thousands of pounds but thousands of tons of fish at sea, explained William Warner.

The *Fairtry* was different from the generations of fishing vessels that preceded her. A stern ramp, or slipway, allowed the boat to haul nets over the stern rather than the side, negating the need for the vessel to be broadside to the swells. A below-deck factory section was equipped with assembly-line machines to gut and fillet fish at high speed and efficiency. A refrigeration plant operated on ammonia or Freon to quickly freeze the catch. And specialized equipment created fishmeal out of processing waste from the machines and unwanted "trash" fish that were also caught in the nets.

The Soviets launched their own version, the *Pushkin*, in 1956. The Pushkin was joined by a sister ship, the *Sverdlovsk*, later that year. By1965, there were 106 Soviet factory trawlers that both caught and processed fish, thirty factory or "mother ships" that only processed, and 425 side trawlers that only fished the North Atlantic from Greenland to Georges Bank.

William Warner, author of the Pulitzer Prize-winning book *Beautiful Swimmers: Watermen, Crabs and the Chesapeake Bay*, was in Newfoundland in the spring of 1976 when he saw his first factory trawler up close. It was a Russian vessel. Its size so impressed him that he wanted to go on board to see the vessel up close, but he was not granted permission at that time. Since the ships had been plying the North Atlantic for over a decade, he knew what the vessel was capable of. Warner realized that due to changing ocean rights because of the treaties and ongoing discussions about fisheries, the factory trawlers that had vacuumed the sea would soon be out of business. But he believed the activity should be documented while they were still operating. He received permission to accompany the ships from five nations, all of which played a part in the fishing drama at sea: the United States, Great Britain, Russia, Spain, and West Germany. His account became the book *Distant Water: The Fate of the North Atlantic Fisherman.*

The Russians were the first foreign nation on Georges Bank with the huge ships, and they were "sweeping up and down Georges with their ships paced out in long diagonal lines, plowing the best fishing grounds like disc harrows in a field," Warner wrote. The total catch of all species for that year was 886,000 tons, which far exceeded the harvests by France, Portugal, and Spain. Nearly a decade later, the Soviet launched a fleet of 710 factory trawlers, 103 mother ships, 2,800 side trawlers, and an armada of supply and support vessels.

Factory trawlers changed the game. Warner suggested that the new rules "meant that for the first time in history fishing vessels of one nation might steam to the shores of any other in the world within two to three weeks,

**An enormous foreign factory ship flanked by huge trawlers with combined capability to catch and process thousands of tons of fish at sea. (Courtesy of National Oceanic and Atmospheric Administration)**

remain there indefinitely with either crew or vessel replacements at sea, fish in all but the worst weather, and be rewarded by an hourly catch rate that surpassed the best effort of conventional side trawlers by twenty-five to fifty percent."

West Germans developed and perfected advancements in net design and a power block to haul the net in 1969. This allowed for midwater trawling so that more ocean could be fished – not just the bottom. Vast schools of herring and other midwater, or pelagic, fish could be harvested, and processing at sea meant the factory trawlers could do it all onsite. The West Germans also added perfected electronics, including four fish finders and eight navigational aids, to their array of equipment. In short, the fish had no chance to escape.

The Spanish also took midwater trawlers to a new level. They had smaller boats, but they developed pair trawling, in which twin ships pulled a net between them as they traveled side by side, separated by a half mile of ocean. The nets were so big that they could bring on board an astonishing 180 tons of fish in one haul, which means that it didn't take many pair trawlers working in an area to nearly empty a bank of fish. These were fish near the bottom of the food chain, fish on which every other fish depended – herring, both sea herring and the alewives that swim up the runs in the spawning rivers on the Cape and elsewhere along the coast; capelin, the fish that the Newfoundlanders depended on to feed the cod for the fish flakes; sardines, anchovies, and mackerel.

Warner was awed by his experiences on the various ships hailing from different countries. "One thinks immediately of the fish," he said. "How any schools can escape the hidden eyes of this armada, since every ship in the fleet is equipped with multiple underwater electronic fish-finders, is difficult to imagine." When the Norwegians, with their centuries of fishing expertise, and Eastern Bloc countries were added to the mix of nations, the results were staggering. It couldn't last.

Long before the factory ships showed up off the U.S. coast, increased fishing efficiency caused concern about fish abundance. In 1871, the U.S. Congress created a Fish Commission because of declines in fish harvests. A fishery research station was opened in Woods Hole four years later. North of the border, Canada's Board of Management established its first marine biological research station at Saint Andrews, New Brunswick. Both countries developed national programs and departments to manage fisheries and territorial waters.

By 1902, eight countries – Denmark, Finland, Germany, the Netherlands, Norway, Sweden, Russia, and the United Kingdom – had formed the International Council for Exploration of the Seas (ICES) to address their unease about the effects of harvesting on fish populations. The U.S. and Canada joined in 1912 and 1967, respectively. ICES focused on research rather than management and continues to be a primary source for international collaborations used by managers.

Countries on both sides of the Atlantic continued to have qualms. In 1946, the charter creating the International Convention for the Northwest Atlantic Fisheries (ICNAF) was signed in Washington, D.C. This agreement pertained to managing the fisheries of the northwest Atlantic outside the territorial waters of eleven coastal countries: Canada, Denmark (representing the Faroe Islands and Greenland), France, Iceland, Italy, Newfoundland, Norway, Portugal, Spain, the United Kingdom, and the United States. FAO and ICES attended as observers.

Thus, three separate groups were all involved in one way or another with fisheries: the United Nations for determining jurisdictional boundaries and providing guidance for fishing in international waters; the International Council for Exploration of the Seas for international research collaborations; and the International Convention for the North Atlantic Fisheries for management in international waters.

With the factory ships, though, the changes in the catch were massive, the competition ceaseless. The mandate to ICNAF commissioners was to make the "maximum sustainable level" of harvest possible and maintain it, but the phrase "maximum sustainable level" was not defined. The common interpretation was to obtain maximum yield in total weight of fish, regardless of species. But when fish were lumped together, it was difficult to know what was happening to individual species. By 1964, when factory fleets were fishing extensively in the northwest Atlantic, Georges Bank and the Gulf of Maine were severely impacted. There, the reported catch skyrocketed from two hundred thousand tons to more than 750,000 tons between 1960 and 1965 and kept growing. In one year alone, the foreign fleet fishing U.S. and Canadian waters topped one thousand vessels and upped its catch to more than two million tons, three times the Canadian harvest and ten times that of the New England fishermen, stated Warner. ICNAF recognized the difficulty in reducing fishing effort and enacted regulations to achieve the objective through conservation measures such as closed areas, protected spawning stocks, mesh sizes, and equipment restrictions.

Enforcement of ICNAF regulations was up to the member states. U.S. and Canadian enforcement agencies could board vessels from their own countries fishing in ICNAF-managed areas as well as those from other member countries. A wrinkle, however, was that prosecuting accused violators was the right of the offending vessel's flag nation.

By the early 1970s, challenges to conservation methods, enforcement, and the crash of some ground-fish populations led to an important change in the concept of maximum sustainable harvest. ICNAF realized there were benefits from not harvesting to maximum sustainable level. In twenty years, the offshore fish factory cities had taken over thirty-six billion tons of fish of multiple species in North American waters. They took everything without utilizing any standards. Size and species made no difference. Juvenile fish were added to the onboard processing plants. Monitoring what was happening in international or national coastal waters was difficult at best. Though regulations were in place, the industry was essentially unhindered.

There had been fair warning of what was to come. Decades earlier, Icelanders had witnessed what happened to the resources around their severely overfished waters, and in 1958 Iceland declared that its territorial waters extended twelve miles from its shores, where foreigners were not allowed to fish. They extended the distance to fifty miles in 1972, and they further expanded their territorial waters to two hundred miles offshore just three years later, creating quite a stir. But Iceland is a small country.

The United States and Canada followed suit, and when they did, the size of the territorial waters they claimed was shocking to the international community. In April 1976, the U.S. exercised its sovereign rights to enact a two hundred-mile exclusive economic zone along its shores. Canada did the same the next year. Other nations followed suit, and in 1979 ICNAF was officially dissolved, replaced by a new multilateral organization, the Northwest Atlantic Fisheries Organization, (NAFO) currently in effect.

With the two hundred-mile EEZ, both the U.S. and Canada thought they were home free, that they could manage their resources for their own people, that they could enjoy a bonanza without foreign interference, and that the mighty cod would be restored to its former abundance. They were wrong.

As each country invested in its fishing industry – with subsidies to Canadian fishermen and incentives by the United States to buy more boats – their domestic harvest effort increased dramatically. New England's trawl fishery swelled from 825 vessels to fourteen hundred in five years, and the bottom gillnet fishery expanded tremendously. Employment in Canada's fishing industry more than doubled during the same period. The combination of additional boats and employment in both countries was seen as a boon to domestic fish production, with domestic economic growth driven by home-grown policy changes, not the foreign factory trawlers fishing in territorial waters. In the long run, though, both policies drove overfishing.

It was a deadly choice. With the increased economic effort came increased but unsustainably high catches, and fish populations could not reproduce within the time or in the numbers necessary to continue to supply that demand. Fishermen and their families, corporate and individual vessel owners and operators, ancillary fishing businesses, and those who depended on fishing to sustain communities saw the increased landings as a sign of economic stability, fueling even more fishing. They all expected that there would be enough fish to keep the industry viable. Without fishing, a lot of people would be unemployed and hurting.

But without fish, there could be no widespread industry. And the cod population was in trouble.

In fifteen years of unimpeded fishing by foreign fleets, the factory mother ships took about eight million tons of cod alone out of the northwestern Atlantic, mostly in Newfoundland and Labrador waters. This is roughly the same amount of cod that was caught by traditional methods in a hundred years from 1650 to 1750.

Cod reach reproductive maturity in three to eight years and may live for twenty-five years. Overfishing devastated the reproduction forces of major stocks of cod. The larger fish with the greatest reproductive potential were removed, and smaller fish with less reproductive potential became the norm. The number of spawning fish declined so rapidly and by such huge amounts that the population crashed. At the rate of harvest, the population could not rebound.

Population dynamics do not depend solely on harvest. That is only one factor. The number, size, and age of spawning fish are equally, if not more, important. But catch statistics offer a glimpse and some perspective on what was happening to the cod fishery. In 1960, New England cod landings totaled sixteen thousand metric tons; by 1976, the annual catch had risen to twenty-five thousand metric tons. In 1977, when the two hundred-mile limit was imposed, the annual harvest was still high at thirty-five thousand metric tons.

During the 15 years of the foreign fleets, from 1960-76, a total of 359 thousand metric tons of cod were taken. In the twenty-five years after the two hundred-mile limit was imposed, a whopping 752 thousand metric tons were fished from New England, indicating that the extra effort by domestic fleets did just as much damage to the stock's sustainability. By 2001, the yearly landings had dropped to fifteen thousand metric tons, and by 2015, the total harvest was merely one and a half thousand metric tons, a sad commentary on the highly restricted fishery.

The story from 1977 to the present demonstrates the biological, social, economic, political, and emotional toll on people and the fish in the Northeast and why the origin of fresh Atlantic cod in U.S. supermarkets often indicates it is not a U.S. product.

Part of the Magnuson-Stevens Fishery Conservation and Management Act that created the two hundred-mile limit for the United States required the formation of management councils to develop the federal rules and procedures under which the fishing industry would operate. The law compelled the councils to consider the biological realities for many target species and the social and economic consequences of management actions in species-specific plans. Thus fish, fishermen, and fishing communities were all part of the process. Some species of groundfish, but not all, were seriously depleted but there was hope that stocks of cod, haddock, flounders, soles, and their relatives would increase with prudent management.

Some wondered if fisheries management and socioeconomic considerations were mutually exclusive or what solution could balance these needs. The goal for each plan was subject to public scrutiny and became the basis of discussion and often acrimonious debate.

A fundamental question was whether fishing effort could be sustained at a level to avoid overfishing without further reducing existing damage to the stocks. If so, could fishing effort be regulated without excluding some segment of the fishing industry or the larger community – or both?

With all these voices and concerns, the New England Fishery Management Council had its hands full from its inception. To rebuild depleted stocks, managers argued that fish populations could rebound only through decreasing fishing effort and protecting known important habitats, such as spawning areas that provided fish a sanctuary from harvest pressure.

The fishermen were torn in different directions, too. Groups that might have been cohesive, splintered. Small-boat owners, like those in Chatham, had interests and needs and a proportionately smaller voice at the meetings, but they challenged the interests of large-boat operators. Technique added to the divisiveness as fishermen who relied on hooks challenged those who used gillnets, and then they teamed up against those with bottom trawls. It was not just owners and operators against the corporations; it was also fisherman against fisherman and community against community.

Many said the government had a regulatory chokehold on the industry and that the fishing culture would change heartlessly and forever. Others countered that fishermen, accustomed to centuries of fishing traditions and a sense of entitlement to take whatever they and the market desired, could not adapt to the new realities. From the fishermen's point of view, whether through paperwork, regulation, enforcement, or debate, any perceived freedom to fish as they wanted, as they had done before, was gone.

Managers needed to know how many fish there were throughout the exclusive economic zone and used the tool of stock assessments. The assessments are surveys to determine numbers of fish by species, size, gender, and numerous other factors. They utilized random stratified sampling in areas of high fish concentrations acknowledged by fishermen as well as in areas not necessarily fished. The National Marine Fisheries Service and the Atlantic States Marine Fisheries Commission collaborated on the assessments, management suggestions, and regulations. Maintaining consistency between federal and state regulations was important to reduce confusion and enforcement difficulties because fish cross boundaries.

Fishermen frequently asserted that there were more fish than the assessments indicated. Yet, the stock assessments were the standard used to develop a regulatory framework. Reliance on them bred widespread mistrust regarding the conclusions from the basic science and led to contentious disagreements between the fishermen and the fishery scientists and managers at both the state and federal levels.

Managers questioned how to best reduce fishing effort. Reducing the fleet was proposed – in essence, laying off a massive number of fishing crews with ripple effects in their close-knit communities. Some suggested buyback programs for the value placed on boats and gear. No one knew whether the proposal would create social good, or how it would affect fishing communities, or if people would remain in the industry – or even continue to live in the region.

Everything was on the table: the boundaries of open areas; when an area could be exempt or open and when harvesting would be prohibited; harvest quantity; appropriate methods to protect spawning stock or net mesh size to protect juvenile fish; fishing bans for specified spawning areas and seasons, limiting the number of days at sea; or initiating quotas for each species. Finally, as the range of proposed regulations became ever more restrictive, how far should they go? No one knew if the restrictions were the tipping point that would bring stocks back or completely destroy the industry. Scientists could not answer the questions with a comfortable degree of certainty, and that fueled fishermen's distrust of their assessments.

The regulations meant enforcement, too. Enforcement meant eyes on the water – a system of onboard observers and a paper trail – to track what happened where. Fishermen were required to file detailed logbooks of where they fished and what they caught. Their records were compared to those kept by dealers, who also had to file details of who brought in what and from where. For people who had difficulty enough battling weather and constant challenges offshore, not to mention the profitability of the fish they did catch, the increased paperwork was another source of contention.

One question kept cropping up throughout the debate. Was it even possible, let alone prudent, to manage the fisheries species by species without accounting for the overall ecology of a place like the Gulf of Maine or Georges Bank? The bottom line, so to speak, was whether fishing for only one species could negatively impact another important species. The answer was unknown; this was new territory for all of the parties.

The combination of management tools – mesh size, closed areas, quotas, limited days at sea – fell short of entirely closing fisheries for any specific species, but they nearly accomplished the same thing.

The regulations also caused what many considered to be a dangerous situation, whether fishermen should take chances with weather and unsafe conditions to fill a maximum allowable quota. A decreased number of allowable days at sea forced them to go out when they normally would not take the risk, a decision that jeopardized their lives. The rules also exacerbated the wasteful practice of catching unwanted fish – bycatch – or exceeding limits of specific species mixed in with legal catches. That meant fish were thrown overboard, returned to the sea – dead. Nobody wanted that outcome.

The restrictive management plans, especially those for the coveted ground-fish resources, resulted in gutting a centuries-old fishing industry dependent on several major species. The effects of the current regulatory climate can be seen in every fishing port in New England. They may be most dramatic in Provincetown, a community tied to the sea like no other on the Cape.

With its beautiful, deep, naturally-protected harbor and close proximity to fishing grounds, Provincetown turned to the sea early on for its economic and social lifeblood. Judy Dutra's 2011 *Nautical Twilight: The Story of a Cape Cod Fishing Family* is a memoir of one Provincetown fishing family. She tells the story of a community that has morphed from a robust fishing town with a Portuguese culture, to a fractured, more diverse but less cohesive community, dependent almost exclusively on tourism instead of fishing.

The Provincetown boats were not longliners. They were draggers that towed nets along the bottom to catch multiple species of fish. "Here in Provincetown, we are losing the last of our kind," Dutra wrote.

"The Eastern-rigged dragger is becoming extinct. I was walking on the wharf behind a man and woman not long ago and I heard the man say, 'I remember when this wharf was filled with fishing boats tied to the pier, four and five deep. I remember when the trucks hauling fish off the pier numbered fifteen in a day. ... Now I hear they are closing the high school because there are no families left in town. I know that fishing families filled the schools when I lived here.' They walked away, his words ringing in my ears. I wanted to cry."

She continued:

"The latest in the thirty-five-year history of the Magnuson/Stevens Act is Amendment 16, Framework 42 of the Small Equity Compliance Guide. It is so restrictive that it is impossible to continue to make a living at fishing. Under these new regulations if our *Richard and Arnold* does not join a sector and wants to fish in the Gulf of Maine, our boat will have 16 days

**Provincetown dragger *Richard and Arnold* owned by the Dutras and the subject of Judy Dutra's 2011 book, *Nautical Twilight: The Story of a Cape Cod Fishing Family*. (Courtesy of Provincetown History Preservation Project)**

to fish. Each day we are allowed 100 pounds of cod fish, 250 pounds of yellowtail flounder, a smattering of by-catch and no flounder at all, we'll have to throw them overboard, dead. Someone told me that the government has no heart, has no soul and that no one is responsible. From what I have witnessed, I am beginning to think that might be true."

In talking to Cape fishermen, the personal and economic effects mix into one tale. Fishing in the twenty-first century is nothing like it was when Tiggie Peluso began fishing out of Chatham and Harwich a couple of generations ago, in the 1940s. Gone are the tremendous schools of fish of every species and the run of cod that the Nauset and Chatham Harbor men fished for on the back side of the Cape in the spring. Gone is the ability to bring home as much fish as a boat will hold. Four thousand-pound catches for a day of fishing are a memory. Gone are the days of "high liners" who caught the most fish consistently or on a lucky day, the camaraderie on the docks with its bantering and jokes, and the envy felt by those who did not succeed that day. Gone are the old ways so well known by fishing families – the wooden boats; the reliance on sounding leads; and the navigational skills, without electronics, that were ingrained in every captain and that he taught his sons and they taught their sons. On the Cape, men who grew up expecting to make their living from the sea did so – for a while; then they turned to carpentry, earthmoving, or other trades or businesses that served the growing influx of tourists and new residents.

In their place are new fishermen, willing to still give it a go as long as they can be on the water. They go to sea in fiberglass or steel boats, eyes glued to the screens of their electronic gadgets on which they are completely reliant. These men are forced to fish no more than an allotted number of days to catch an allotted amount of fish. There is no choice. Before they leave port, fishermen must notify the National Marine Fisheries Service of their intent, what they will be fishing for and where, and then report their catch – the amount, species, and location – on their return.

Instead of setting long lines of hooks or towing a net or setting a gillnet, many fishermen are turning to another resource: setting hundreds of lobster pots in places where cod once flourished. Scientists believe the decline of cod, a predator, has been a boon for lobsters. By the summer of 2012, the booming lobster population had driven consumer prices very low, so lobstermen now trap for greater quantity just to cover the cost of a trip out of the harbor. One wonders how long the rate of harvest will continue and if the lobster populations will be sustainable with current regulations and practices.

And yet those lobster pot lines may entangle the highly-endangered and federally-protected North Atlantic right whale, and researchers and fishermen are searching together for ways to fish effectively while safeguarding the whales. Fisheries have become increasingly intertwined with issues of protecting other inhabitants of the same sea.

The cod stocks also collapsed in Newfoundland. Technological advances, inherent uncertainty about stock assessments, and the sociopolitical climate were all factors. The country chose to save the fish, even with its insight into how such regulations would impact the fishing culture.

The Canadian Ministry of Fisheries and Oceans imposed a moratorium on harvesting cod in 1992, intended for a two-year duration. It came too late. The spawning biomass had already crashed by 75 to 90 percent for five of the six major regional populations. Within a few years after the ban, the population of northern cod was estimated to be a mere 1,870 tons and still dropping.

The Newfoundland story became a dreadful example of overfishing to the breaking point. Some experts at the time suggested that the reproductively healthy stock might not recover to the point where it could stand on its own for fifteen or more years in the face of industrial fishing of any kind. It has been well over fifteen years, and that time has not yet arrived.

With limited exceptions, the Newfoundland cod fishery shuttered and is defunct. Hardly a visible trace remains of that once-thriving, centuries-old industry. Over thirty thousand Newfoundland and Labrador residents lost their livelihood. The fishery did not rebound. The ecosystem had altered, and no one knew if it had been altered irreparably. Invertebrates such as shrimp and crabs increased dramatically in the absence of the predaceous cod. Capelin, a major food source for cod, were not there.

But finally, more than twenty five years after the Canadian ban, there are some hopeful signs. The ecosystem appears to be changing to one favorable to capelin. Meanwhile, the cod harvest prohibition continues.

When – or if – the cod and other fish recover, what lessons will we have learned? One thing this example illustrates is that nature does not live by our timetable. Ecological recovery can often be gradual – decades, or generations, in the making.

As in Newfoundland, the fishing industry on Cape Cod is a shadow of the past. It has become a different world below the surface of the Atlantic. The loss of the fishing industry has also meant a different world on land. Cape Cod has become a tourist haven and former working waterfronts have become expensive real estate for non-water-dependent businesses like restaurants and non-commercial land for vacation homes.

The future of cod around Cape Cod is not secure. Uncertainty in almost every facet of cod management has resulted in rampant distrust, rancor, and animosity. And now there is a new looming wrinkle never encountered before – climate change and the warming of the seas. With warmer seas, how will the ecology of the oceans around Cape Cod manifest that temperature differential, and will cod be a part of that new ecology? Fisheries scientists have seen a decrease in zooplankton species at the base of the food chain that are important for juvenile cod and have historically been abundant in traditional cod spawning areas. Climate change has been cited as a reason for the decrease of the plankton. The unprecedented ecological upheaval we are experiencing now, and the future we are on the way to experiencing, present a quantum leap in uncertainty for the fate of a fish that thrives in cold water.

Cape fishermen remain hopeful despite the uncertainty about their industry. They are adapting to the lack of cod as they always have, by finding alternative species or methods and by developing partnerships in an industry in which independence has been the traditional way.

The Cape Cod Commercial Fishermen's Alliance is working diligently to make sure that the industry survives. Based in Chatham but representing the entire Cape, the alliance has been in the forefront of industry adaptation. One species garnering its attention is the sustainable dogfish, a type of small shark. Currently, most of the catch is exported to the UK, but the alliance is working with partners to gain U.S. market acceptance. Marketed as "Cape shark," it is a mild white fish, not flaky, with a texture more akin to Mahi Mahi.

The alliance also purchases expensive licenses within an Individual Transferrable Quota system to fish for certain species. It's known as sector management. Small-boat owners would not be able to purchase those licenses themselves. The alliance then leases or sells them on an open market, a system that allows these men to fish. The alliance has been politically active and lends its voice to discussions at the New England Management Council and the Atlantic States Marine Fisheries Commission, sharing their vision for "Small Boats. Big Ideas." By banding together, they have found strength for their organization by including all types of offshore and inshore fisheries.

So questions remain, especially for cod. Will the management plans and regulations that severely limit harvest help species to rebound at some point? If yes, will there be fishermen and the infrastructure they require to take advantage of the new bounty? Will a new fishery be sustainable?

Will we have learned the lesson, or will history repeat itself yet again?

Is it possible to have it all? Or, in our quest to have it all, is it possible that we have almost annihilated the fish that started it all?

An even bigger question is whether we will come to understand enough about the ocean's mysteries to integrate its complexities into a plan for the entire ecosystem without sacrificing one piece for another.

We do not know the answers to any of these questions at this point. As we continue scientific investigations, we must also listen to those out on the water as they share their own astute observations. Above all, we need to make sure that whatever our actions may be, they do not cause irreparable harm to the environment.

Chapter 16

# Shifting Baselines – Shellfish and Estuaries

As winter warms into spring, the coastal banks along the seashores burst into bloom. White blossoms decorate beach plums before they leaf out, while golden scotch broom complements the perennial deep green of that shrub. Meanwhile, the bays come alive. Light skin-toned, nickel-sized jellyfish drift with the tides, growing steadily until they are five or six inches across. A sudden swirl of water heralds "schoolies," small striped bass chasing smaller fish. The bass arrive when the water warms to nearly fifty degrees, coinciding with the appearance of herring heading to the creeks that lead upstream to the freshwater ponds. Gulls, attuned to the bounty, squawk overhead, then congregate to claim their share of the migrating herring.

The sandy bottom that was so visible in the winter, when every ripple, shell, hole, and creature moving along it could be seen, becomes more obscure as plankton clouds the water and the first green eelgrass leaves emerge. Soon, the individual eelgrass plants with their many leaves and seedpods merge together to form dense underwater grass meadows. The eelgrass hides pipefish, relatives of the seahorse, and many other fish, spider crabs with their rounded bodies and long legs, mud crabs, snails, and whelks that stealthily wander through the meadow searching for a meal.

As the bright green shoots of the saltmarsh grass peek through the tawny brown detritus of the last season's growth to find the sunlight, the marsh creeks fill with life. Crabs emerge from their winter burrows, while baitfish hatch and grow quickly.

The shorebirds arrive at about the same time the bass arrive, or maybe a bit earlier. Piping plovers, a threatened species that nests on beaches, pair off to begin their families. Laying clutches of four almost perfectly camouflaged eggs, they sit on the nests until their little sand-colored fluff balls hatch. As the little ones grow, they run down to the water's edge to feed on miniscule invertebrates. Highly vulnerable to late-season storms that wash away their nests, the birds are also threatened because they are preyed upon by foxes, coyotes, skunks, raccoons, some birds, and other predators. They are also susceptible to man's activities. Off-road vehicles driven on the beach leave deep ruts in the sand that the tiny birds cannot avoid, trapping them in what must seem like deep chasms. Dogs on the beach and kites in the air disturb the plover parents who leave their young to try to ward off the danger. Beach management now takes the plovers' life requirements into account, and beaches are often closed to vehicular and pedestrian traffic where the plovers are most vulnerable. Unfortunately, the closures occur during the most popular time of year for the beach-going public, infuriating many beachgoers and subjecting beach managers and employees to verbal abuse. Bumper stickers proclaim "Piping Plovers Taste Like Chicken."

Plovers are only one species. Shorebirds begin to arrive in the spring and stop to feed and rest on the way to their nesting areas far to the north. Terns, bright-orange-billed oystercatchers, lesser and greater yellowlegs, white and grey sandpipers, sanderlings, godwits, and dowitchers all feed on the beaches and intertidal flats of the estuaries.

Horseshoe crabs, a relative of the spider rather than crab, also inhabit the estuaries. They have existed, essentially unchanged, for hundreds of millions of years. In the spring, the large females crawl along the sandy bottom, lugging at least one male attached to their shells, and usually two or three males stacked on top of one another and her. Their pointed, three-sided saber-tapered tails leave a distinctive trail in the sand as the female hauls her suitors to a high patch of beach. The stacked crabs begin their egg-laying ritual on the highest spring moon tides, the females laying hundreds of eggs in the upper reaches of the intertidal beaches where shorebirds like the red knot, one of the chief predators, gorge on the eggs. An infrequent visitor to the Cape Cod region, the red knot is the dominant shorebird predator in the mid-Atlantic region, particularly in New Jersey. Of the eggs that hatch, the hatchlings go through many shell castings as they grow, the cast-off perfectly formed shells floating in the waters and end up on the beaches.

In nature, often what was once the prey, such as the horseshoe crab eggs, becomes the predator. Horseshoe crabs are shellfish predators but do not feed by crushing food with their claws like conventional crabs. Rather, they use the mechanical advantage of their jointed legs to crush small prey near their mouth. The brittle shells of juvenile clams are easy pickings.

Life in the bays ratchets up a few notches in the summer for fish and the many species of invertebrates. We humans pay little or no attention to most invertebrate species, but they are important in the web of life. We do pay attention to shellfish, although only five species are the most important commercially – softshell and hardshell clams, oysters, mussels, and scallops. In summer, demand is high as tourists flock to the seashore to savor their oysters or littlenecks on the half shell, or plates of steamers or fried clams, or mussels marinara, or fried scallops.

Softshell clams and mussels began to spew their millions of eggs and sperm into the water in the early spring when a chance meeting of egg and sperm begin the life process. Quahaugs, scallops, and oysters need warmer water and begin the process in late spring, through the summer, and into early fall. Two days after fertilization, the veligers, microscopic larvae, develop a swimming organ – though "swimming" may not be the appropriate term, as they drift wherever the tides and currents take them. Their swimming is mostly a means to keep them moving, though in laboratories they have been observed moving up and down with some regularity. As a member of the zooplankton community for these two early weeks of their lifecycle, in which a single drop of water may contain dozens of microscopic species, they are food for just about anything that takes in water. Once they metamorphose and acquire their hard shells, they grow quickly, feeding in turn on the huge amount of plankton in the water.

Larger striped bass and hungry bluefish arrive in the bays to feed on sand lance and smaller fish, while crabs forage for any morsel. Odd-colored, textured, or shaped egg cases abound. Worms form in amorphous gelatinous masses, whelks leave a string of beige quarter-sized disks strung together on the bottom like a twisted necklace, and small snails cement miniature bowling pin-shaped egg sacs to shell fragments.

All sorts of new generations are everywhere. Ospreys, birds of prey that are magnificent fishers, have returned after their near demise because of the now-banned pesticide DDT. They build intricate and large nests, often atop high platforms made specifically for them by conservation groups. Osprey parents take turns protecting the eggs or hatchlings in their nests above the surrounding marsh while the partner parents fish for food.

By September, the estuaries begin to change subtly. July's hazy heat gives way to spectacularly clear and vibrant days, when the water is warm enough for swimming and the nights are cool enough to sleep soundly. The strong southwest wind of summer afternoons switches more often to a northerly direction. Young ospreys have fledged and learned to fish for themselves as they prepare for a long flight south. Waves of shorebirds arrive, pecking away at sandy intertidal flats to "stock up" for their own southern migrations. The flats harbor hundreds of interstitial species, those that live between the sand grains, unseen and unknown to most of us, but vital to the birds. Herring left the freshwater ponds and estuaries months ago to head back to sea, and now their progeny migrate to the sea as well. Bass and blues begin their journey to warmer waters. Everything it seems, is on the move, while the green of the summer marsh and dunes transforms to the golden seed pods of autumn, imparting a multihued tint to the land at the edge of the sea.

By the end of the fall, the shorebirds are gone, replaced by ducks and geese. Any waterborne species that survived the summer and fall begin biological preparations for winter. As the temperature drops, metabolism slows. Those that survive the winter have stored enough energy to live until the water temperature again rises. It is the cycle of life in the estuaries, and there are always winners and losers. Predators and prey determine the rules of the game.

Estuaries, where land and sea meet, have discreet habitats. Fringing the upland are coastal banks, marshes, beaches, tidal flats, rocky shores, and cobble beaches. There are submerged lands, overlying waters, and transitions from one to another, each with its own requirements. There are barrier beach habitats of tidal sandflats, dunes, and the great beach or massive rock headlands, and offshore ledges that protect the estuaries from the ravages of the Atlantic Ocean beyond.

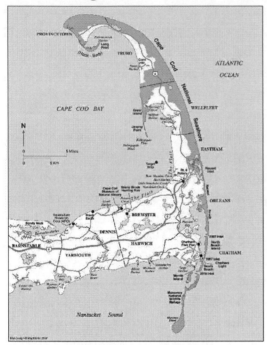

**Map of Outer Cape indicates geographic perspective of area. (Mapworks)**

The coastal banks may be vegetated or bare sand and clay – the latter a remainder of a storm that chewed away at the base of the bank, causing a landslide from above. Saltmarshes fringe the estuaries, interspersed with sandy or cobbled beaches. High tide covers the marshes, while low tide exposes intertidal flats, often a hard sandy bottom, sometimes a mucky, messy one. The sediment turns increasingly silty in the deeper water. And where there is less turbulence on the bottom, the sediment is soft, a black, tar-like muck. Even the muck is a habitat, though not a very healthy one; only the most opportunistic species can survive there, like the red pencil-lead-wide polychaete worms. The overlying water is a habitat for everything that swims or moves through the watery world. There may also be islands with more beaches, marshes, thickets, and upland, all perfect habitats for egrets, pure-white, long-necked summer visitors – birds more commonly associated with the tropical areas where they are year-round residents rather than Cape Cod.

In the estuaries, freshwater from the land – creeks, streams, rivers, and groundwater – and saltwater from the sea combine to create the perfect circumstances for this tremendous diversity of life to thrive.

Over one-third of Orleans is salt water. Getting to know what shellfish were located where in the three separate estuaries of Cape Cod Bay, Nauset, and Pleasant Bay was instructive for my role as the biologist for Orleans' Shellfish Department. Whether on foot or drifting in a boat in shallow water, watching creatures scuttle along the bottom or swim in their watery home, I gained an inkling of the marine world's wonders.

**Pre-1987 aerial view of Orleans looking south prior to Chatham breach. Nauset estuary is in foreground, Pleasant Bay toward the top and Cape Cod Bay on the right. (Courtesy of Kelsey-Kennard Photography)**

On the expansive flats of Cape Cod Bay, fanning out from the shore, jingle, slipper, and razor clam shells were the most common. All three were aptly named: jingle shells "tinkled" when strung up as a set of hanging chimes; slipper shells, also called boat shells, resembled rounded-soled slippers; and razor clams were reminiscent of old-fashioned straight razors and nearly as sharp, as any barefooted person cut by one could attest. The egg cases of two animals were on the sand too – infrequent single dried black leathery skate pouches, and beige wafer-shaped disk whelk "necklaces." Rippled sand was a sign of an unstable bottom, due to wave action, where clams and quahaugs were unlikely to set. The bars were interspersed by a few never-dry tide pools in depressions, where balloon-like gelatinous worm egg masses several inches wide billowed from holes in the sand, joined by periwinkles, a few crabs, and baitfish. Some pools could be deceptively deep, as the tide flooded them first. While clams or quahaugs were nearly absent, some moon snails, predators of the clams, were foraging for something to eat. The mile wide ribbon of flats appeared nearly barren.

At Nauset, on the other hand, there was tremendous variety of marine life: horseshoe crabs and crustaceans (lobsters, green crabs, rock crabs) and filter-feeding bivalves (clams, quahaugs, mussels, and sometimes scallops). There were no oysters because they have been absent for generations. But tiny gem clams, mini-versions of quahaugs, except with purple on the outside of the shell instead of on the inside, were abundant. Deposit-feeding bivalves like Macomas, which are found in muddy areas and gain nourishment from both silt and the plankton that they take in, were there as well as single-shelled mollusks – moon snails, mud snails, and periwinkles. There were worms of many widths and lengths, from long flat "ribbons" to the type typically used for bass or flounder bait. It was obvious that Nauset was much more productive than the flats of Cape Cod Bay.

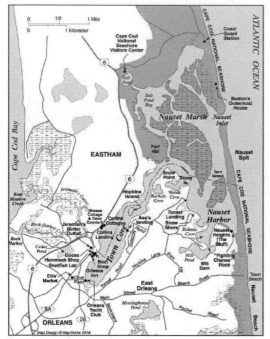

**Map of Nauset Estuary. Fort Hill is located in the center of the map. (Mapworks)**

Pleasant Bay, much larger than Nauset, is shared by three towns, but Orleans' portion, in the upper part of the bay, is the largest. Lobsters, mussels, and moon snails were not in Orleans, but they were in the Chatham portion of the bay nearer to the inlet. Blue crabs and mantis shrimp were in the bay that I didn't see in the other two estuaries. Horseshoe crabs were abundant, but green crabs were not as numerous as in Nauset. Clams were sometimes very numerous and sometimes hard to find, and quahaugs were spotty. Scallops were noticeably larger when they were available.

No more than a mile separated Cape Cod Bay from Nauset and Nauset

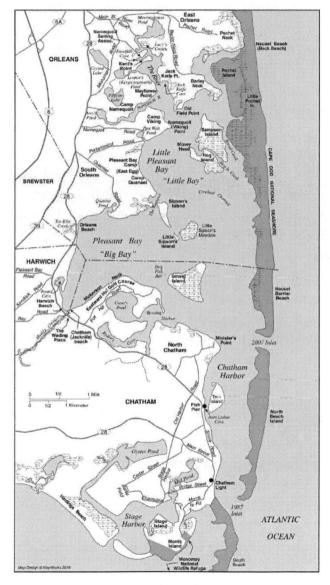

**Map of Pleasant Bay. (Mapworks)**

from Pleasant Bay, yet there were differences in the three separate and distinct estuaries. An explanation was that before the late 1980s, Pleasant Bay was the only one that had water coming in from both the Atlantic Ocean and Nantucket Sound, on the south side of the Cape. That distinction was symbolic of a dividing line, and Orleans represented a much larger picture.

The Cape has been described as a biogeographic boundary for many estuarine species. Cape Cod, appended to the shoreline of Massachusetts, is only forty miles long, and its hook-shaped arm is another thirty miles. Its geographic placement may help to explain the species' patterns. Generally, Cape Cod Bay and the Atlantic Ocean side of the Cape form the southern boundary for northern, or boreal, species that prefer cold water. Nantucket Sound is the northern boundary for species that prefer warmer waters. The Cape blocks the species from mixing. The Cape Cod Canal may provide a conduit for some mingling as tidal water sloshes from Cape Cod Bay to Buzzards Bay and southern waters, but the role of the canal in species distribution is unclear. Despite that uncertainty, the demarcation line is rather distinct between northern species in Cape Cod Bay and southern ones in Nantucket Sound. Yet in Orleans, at the elbow of the Cape, there are species from both zones within the three estuaries.

We can identify many species in and around estuaries, but their interrelationships are more obscure. We tend to pay attention to the species we can see and those that are useful to us. Not much has changed on that score for centuries.

Seventeenth-century chroniclers such as John Smith and William Wood described the marine life they encountered along the Atlantic coast, and their observations remain valuable and detailed inventories.

Smith addressed New England in general, assembling a vast catalogue of sea mammals, fish, and crustaceans: whales, grampus (pilot whales), turbot, sturgeon, cod, hake, haddock, halibut, sole, cusk, ling, shark, mackerel, herring, mullet, bass, cunner, perch, eels, crabs, lobster, mussels, and whelks. He wrote that every river was home to sturgeon and salmon, and a person could take clams on almost any sandy flat and could easily find the ubiquitous lobsters.

Wood in his *New England's Prospect* declared the shoals off Cape Cod as one of the best places for sturgeon. They were twelve to eighteen feet long. He talked of catching halibut "two yards long, one yard wide and a foot thick." Wood described twenty-pound lobsters, oysters a foot long, and clams so numerous that "a man running along these clam banks will presently be made all wet by their spouting of water out of those small holes." Thus, he not only described the species he saw; he also described their size, the time of year when they were present, and to some degree, their abundance. And we have no evidence to the contrary.

Between 1965 and 1972, the Massachusetts Division of Marine Fisheries published surveys of fifteen estuaries in the state, including three for the Cape: Pleasant Bay in 1967, Waquoit Bay in 1971, and Wellfleet Harbor in 1972. Also included were two large Massachusetts Bay estuaries – Plum Island Sound north of Boston in 1968 and Plymouth-Kingston-Duxbury south of Boston in 1974. Although they did not sample for crustaceans or other invertebrates, sixty species of fish and shellfish were identified in Pleasant Bay. It was the first survey of its type and established a baseline for the bay. Through a cooperative agreement, the Pleasant Bay Alliance and Center for Coastal Studies in Provincetown began conducting a similar study in 2015. When completed, it will offer a comparison to determine if the earlier baseline has shifted.

Rutgers University conducted an ecological assessment of the Nauset estuary for the National Park Service in 1989. That assessment found that Nauset was an important spawning area for winter flounder; a nursery area for offshore commercial fish species such as cod, Atlantic tomcod, pollock, and hake, all of which later move offshore; and a feeding area for bass and bluefish. The analysis revealed forty-nine species of fish and eleven species of decapod crustaceans, ten-legged lobsters and crabs, and twenty other crustacean species, thirty-one species of mollusks, or shellfish, and thirty-two species of polychaete worms living in sediment. Also, for the first time, the researchers documented a vastly important discovery – a natural inshore lobster nursery where very small lobsters found refuge from predators. They also reported that forty-seven species of birds utilized the estuary during daylight hours. It was an amazingly full inventory that identified cod, haddock, hake, sole, bass, shark, mackerel, eels, crabs, lobsters, mussels, whelks, and many more species.

The Rutgers survey showed that the nearshore, intertidal, and subtidal areas are not just habitats for clams and quahaugs, but also home to crabs, worms, and other forms of life that provide food for fish and for thousands of birds that stop on their yearly migrations north and south. The water, especially in the summer, is teeming with the plankton and young-of-the-year that will become the food, the fish, the shellfish, the predators, and the prey of the coming year.

Taken together, the Pleasant Bay and Nauset surveys identified species mentioned in the early seventeenth-century lists by Wood, Smith, and others. All of those early chroniclers observed a great many species, but they did not elaborate on the numbers of individuals or on the interconnections between the species that they saw. Their lists serve as baselines, a starting place for the types of marine creatures spotted there. Finding the species present several centuries later establishes a continuum for marine species within the estuaries. However, while modern surveys include estimates of the numbers of individuals of each species whereas the older lists do not, there can be no comparison of population baselines. All we have to go on is Wood's descriptions which indicate far greater abundance than what we see now. Judging by those seventeenth-century descriptions, the species present may be the same today, but the number of individuals may be a completely different story – a hugely shifted biological baseline.

Nauset and Pleasant Bay are examples of shifting physical baselines as well. Samuel de Champlain sailed into Nauset Harbor five centuries ago. He found a harbor about "three to four leagues in circuit," ten miles in circumference, and two fathoms, twelve feet deep at high tide. His pictorial map illustrated the layout – a narrow inlet from the Atlantic leading to a nearly circular harbor, with several rivers or creeks emptying into it and vegetation along the shore.

Champlain's map and his 1605 description form the baseline for our understanding of the historical dimensions that he observed. Today, Nauset Harbor is a bay approximately half the size of what Champlain saw, half of which is an extensive saltmarsh. The water portion has transformed to just one-quarter of the original size and is at least three-quarters shallower. Nauset Harbor is still protected by sand dunes on the eastern side, but the dunes that Champlain described were at least a half mile farther east than they are now and currently under water. (See Champlain's map earlier, Chapter 6 p. 71.)

One of the rivers Champlain depicted, a connection between Pleasant Bay and Nauset, no longer exists. For generations, it served as an inland waterway between Nantucket Sound and Nauset Harbor. With it and Jeremiah's Gutter, Native Americans could travel from the sound to Cape Cod Bay without traversing the treacherous Atlantic. But the creek filled in long ago, and by the mid-nineteenth century it was no longer featured on any maps.

Intertidal sand bars also shift position depending on currents within the harbor. Channels are shallow and tricky, and last year's channel is this year's sand bar. Saltmarshes can form where there were sandflats and flowing water only a few decades earlier. Over time, even the relatively short geologic scale of a person's life, the harbor is very different. Over centuries, it becomes a completely different body of water. Four hundred years have brought so many vital changes to its size, shape, and ecological state that Champlain might not recognize it now.

Physical alterations to the landscape along the ocean are visually apparent. The forces that constantly rework the landscape are subtle on normal days with just the tide at work, but add a storm and the result can be stunning. From decade to decade, these transformations are dramatic. Storms produce wash-overs, places where waves literally crash through the dunes, creating passageways for the water to flow from the ocean side of the barrier beach to the inside bay or harbor. All the sand that was a dune washes into the bay. Over time and many wash-overs, dunes roll over from the sea and migrate west, and the sea encroaches on the land while the harbor gets shallower and less navigable. This movement, called barrier beach dynamics, forms another baseline – a constantly shifting one.

The numerous bays and estuaries of Cape Cod were brimming with shellfish in the seventeenth century. Clams, quahaugs, scallops, mussels, and oysters were all available, although early colonists considered scallop bodies poisonous, and mussels as a food source were rarely, if ever, mentioned in early histories. Wampanoag middens indicate that quahaugs and oysters were those Native Americans' mainstays for centuries. Champlain named Wellfleet Harbor "Port aux Huitres" or Port of Oysters, David Wright has written in *The Famous Beds of Wellfleet: A Shellfishing History.*

Massachusetts was in the forefront of research into the natural histories of four species of shellfish. With only one assistant, Dr. David Belding, a physician, executed an exhaustive investigation from 1905 to 1910 into the life history, ecology, and fishery of quahaugs, scallops, clams, and oysters. He conducted experiments on spawning long before hatchery techniques were common practices. He developed natural "spat catching," or methods to "trap" spat in the wild, to determine rate of growth factors that indicate productive and less productive areas. His remarkable work remains the basis of natural history for these animals. His research was published in monographs by the Massachusetts Fisheries and Game Commission. Reprinted in one volume in 2004 by the Cape Cod Cooperative Extension, *The Works of David L. Belding, M.D., Biologist: Early 20th Century Shellfish Research in Massachusetts; Quahog and Oyster Fishery, The Scallop Fishery, The Soft-Shell Clam Fishery* remains an extremely valuable resource for anyone seeking to understand the biology of the animal as well as historical insight into the fishery of the early twentieth century.

The physical changes to the estuaries are, for the most part, natural changes. The biological changes are not always natural. The estuaries' unique nature of mixing fresh and saltwater, land and sea, makes these magnificent bays vulnerable to changes caused by human use. Estuaries are resilient, but only to a point. We have altered and destroyed habitats, and we have harvested too much of certain species. We have added pollutants. We have shunted storm water that carries numerous contaminants from roads directly to the estuaries. We have not figured out an effective method to handle wastewater from our homes and businesses, and as we struggle to determine the best course of action and debate the merits of possible solutions in an attempt to inch forward, we continually add nutrients that cause the water quality to degrade. We have tipped the balance, and we do not know if our actions have caused irreparable damage. What we do know is that we have created a very different baseline.

We have also developed methods to culture shellfish and have established a vibrant shellfish aquaculture industry. We have created programs that encourage residents to get involved as volunteers and provide public education on the issues. But we do not yet know if these efforts will hold the line against further degradation or provide for habitat improvement and an environment with greater abundance. Public education on all levels has been an effective tool in highlighting the problems, resulting, at the very least, in an acknowledgement of the issues. Developing the public resolve to finance and act on solutions has been more difficult to achieve.

# Chapter 17
# Quahauging

Eben Cummings untied his twenty-eight-foot catboat at the rickety-appearing but deceptively-sturdy dock at Rock Harbor. It was the early 1920s, and while several of Eben's friends still sailed out to the bay as generations had done before them, Eben had altered his boat for newer technology. He had replaced the mast with his one and one-half horsepower, two-cylinder, four-cycle engine. It may have been a day when the dawn was still an hour away as he eased away from the dock and putt-putted his way through the thirty-foot-wide marsh creek to the shallow channel beyond.

The tide was ebbing fast, and soon there would not be enough water to float his boat over the mile and a half of sandflats that separated the harbor from the deeper water offshore. The channel, such as it was, was ten feet deep at high tide but only a few inches at low tide and was marked by fifteen to twenty-foot trees. Fishermen placed the trees each year by pumping them into the sandflats to replace those destroyed by winter ice. With the channel navigable for only two hours before and after high tide, they had two choices: stay out for about eight hours over the low tide, or stay out for up to twelve hours over the high tide and come back as it began to ebb once again.

In the dim predawn light, Eben could still make out landmarks on the shore to find his ranges, lining up a church steeple with a particular house for instance. His ranges told him where he had fished the day before – in the channel near Stony Bar, south of what once was Billingsgate Island. Once the ranges were lined up correctly, he let out the stern anchor, one of two attached to a six hundred-foot line called the anchor rode. When the line was nearly all payed out, he cleated it off and tossed the rest of the line attached to the bow anchor as far as he could. When both anchors were set, the boat was held taut bow and stern, heading into the tide. Then he shut off the engine.

Eben then attached a three-foot-wide, twenty-plus-pound iron bull rake, armed with twenty-five evenly spaced teeth, to one of the thirty, forty, or fifty-foot wooden poles. He carried multiple poles to use whichever pole was necessary for the depth. The rectangular rake was forged with four semicircular hoops. A net was attached to the frame to hold the quahaugs that the teeth dug out of the bay bottom.

Eben stood on the bow of his boat and slid the bull rake down the anchor rode until it hit the bottom. With the tide pushing against the rake, the teeth dug into the sediment and Eben was ready to rake, dislodging the quahaugs from the sand. He stepped slowly aft, balancing precariously on the narrow space between the cockpit coaming and gunwale, a space not wide enough to place his feet side by side. With the pole on his shoulder, he pressed down and lifted slightly with short jerky motions, as he raked through the sand to dislodge the quahaugs buried a few inches below the surface. Foot by foot, he moved slowly sternward until he felt the haul was enough to bring up to the boat. Inching back toward the bow, he pushed the pole to a vertical position until the teeth broke free of the bottom. Then, hand under hand, he brought the heavy rake, laden with thick-shelled quahaugs, empty shells, small rocks and maybe a sea clam or two, aboard the boat for

**Eben Cummings on the bow of his catboat on Cape Cod Bay. (Courtesy of Cape Cod National Seashore)**

culling. He threw the shack – the shells and any other unwanted things caught in the rake – back over the side and bagged the quahaugs into bushel burlap bags.

Eben continued raking as the sky brightened with the light of a new day. He jerked the rake in and out of the sand for hours. Finally, he hoisted the anchors, started his engine, and slowly motored back to the creek at Rock Harbor. The day had been relatively calm, with a breeze out of the southwest. If the wind had been from the northwest, it would have been too choppy to fish safely, and he would have had to ride it out until he could get back to the protection of the creek. But this was a good day for him and his family. He had harvested ten bushels of quahaugs. Warren Darling recounted the laborious process of quahauging in the deep water of Cape Cod Bay in his 1984 book *Quahauging out of Rock Harbor*, 1890-1930.

≈≈≈

Quahaugs are a popular dish. These hard-shell clams are separated in the marketplace by size names: littlenecks, cherrys and chowders but they are all the same animal. The smallest legal-sized ones in Massachusetts are littlenecks. Littlenecks are measured by thickness of the two shells and must be one inch thick. Cherrystones, or cherrys are between two and a half to three and a half inches, measured by the length of the shell in longest diameter and chowders are anything larger than that. Littlenecks and cherrystones are eaten raw as clams casino and are used for cooking. Chowders are used for their namesake and for fritters, and they're also baked and stuffed and used for dishes that require pieces of quahaugs.

New England tribes called the hard-shell clams quahaugs, but the English spellings have varied over the years: quahauag, quahog, and quohog, to name three. The term is useful because it differentiates this species from many other species of clams. The modern marketplace, though, now calls quahaugs clams, an unfortunate change in vernacular. In communities where both quahaugs and soft-shell clams are present, the latter are generally referred to simply as clams. Without the differentiation, someone asking for clams inevitably leads to the second question – which one?

Properly managed, quahaugs can be a reliable, productive fishery. They live a long time – twenty to thirty years or more is not unusual. The oldest recorded age was forty-six until 2011 when researchers analyzed a large shell and determined that the animal was 106 years old at death. Their longevity and resultant great reproductive potential lead to more stable populations than other shellfish, although successful spawning by large older animals requires that everything be just right. Even if they produce a poor set one year they can typically recover as long as they are not overharvested.

Quahauging with a hand rake was a hard way to make a day's pay. The term "bull-raker" was an apt name for the fishermen who tossed around the heavy bull rakes attached to awkward long wooden poles. It took bull strength to rake quahaugs in the deep water of Cape Cod Bay while balancing on the rail of a bobbing catboat. Relatively few men participated in such a labor-intensive fishery. Bull-raking kept the subtidal quahaug population balanced; naturally limiting the amount of quahaugs harvested ensured a continuing supply, wrote Darling.

But the times and methods changed drastically with technology. The internal combustion engine became available for boats, and operating larger and more powerful engines became more affordable, transforming the situation for many fishermen. The ramifications came later.

Draggers, primarily from Wellfleet and Orleans, began to harvest great quantities of quahaugs. They used a knife dredge, locally called a rocking-chair dredge. Quahaugs, with their short siphon that limits how far they can dig down, live only a few inches below the sediment surface. The dredge dug below the quahaugs to lift them out

**Hydraulic dredge used in Cape Cod Bay for harvesting quahaugs. (Photo by author)**

of the sand, and they were dropped into the steel "bag" behind the dredge. The method was hard on boats because it was not a smooth maneuver. The dredge created a sort of jerky, "rocking chair" motion as it plowed through the sand. Later came hydraulic dredging, in which a set of high-pressure water jets were welded to the front of the dredge. As the device was towed along the bottom, the sand was loosened by the high-pressure water, dislodging the quahaugs that were scooped up by the dredge. Boats could harvest more efficiently and without as much stress, and the shellfish population began to become depleted. Once again, technological efficiency led to decreased stocks.

Towns in Massachusetts enact and enforce shellfish regulations for open and closed areas, harvest limits, fees for licenses, and other restrictions. Towns are also responsible for shellfish propagation in their own waters and do everything that time and funds allow to increase the amount of shellfish that are approved by the state for harvest. Many people think the Massachusetts model of individual town shellfish management is ludicrous, if not downright crazy, since most states consolidate these management processes at the state level. However, while regulations vary by community, the amount of propagation and the knowledge about the shellfish habitat within Massachusetts communities often exceeds state-level management capabilities elsewhere, where funding and personnel are limited.

**Phil Schwind, Shellfish Constable for Eastham, transplanting Cape Cod Bay parent stock for town's propagation program. (Phil Schwind collection)**

Propagation has been an important component of municipal shellfish management programs. For years, quahaugs harvested from Cape Cod Bay were used as transplant stock, at least in Orleans and Eastham, while numerous towns bought quahaugs from other parts of the state. The

**Intertidal or subtidal bottom boxes utilizes natural sediment.**
**(Courtesy of Barnstable Natural Resources Department)**

purpose was to plant parent stock that would spawn to produce new generations of shellfish. It was not a proven method, but it was basically all that was available to the towns for this species. In many towns, the areas set aside for transplanting were open to harvest at certain times. Called "put and take," the program worked. Not all of the transplanted quahaugs were harvested, and they lived to spawn.

By the mid-1970s, several communities, including Orleans, were working with very small hatchery seed quahaugs. Shellfish hatcheries are land-based facilities adjacent to a source of saltwater that is pumped into the building. Hatcheries produce seed shellfish beginning with the spawning process. They "condition" brood stock by raising the winter water temperature and feeding the adults algae that they culture to get them to "ripen" out of season. When the animals' gonads are "ripe" and they are ready to spawn, the hatchery operator raises the temperature slowly and then lowers it to control the spawning activity. Raising and lowering the temperature is generally enough to induce the shellfish to spawn. When they spawn, the sex can be determined and males and females are kept apart so that eggs are collected in separate containers. Sperm is added to fertilize the eggs, and after forty-eight hours, the microscopic animals have developed into what are called "D" stage larvae. They develop a swimming organ and remain a swimming animal for about two weeks. Then they metamorphose, gain their hard shell, lose their swimming organ, develop a foot, and resemble tiny versions of the species we recognize later as adults. Hatcheries continually feed and maintain the very small animals and then distribute them to culturists, who use various methods of nursery culture for the next phase of growth.

**Sand-filled rafts for quahaug nursery culture. (Photo by author)**

Culturing millions of seed shellfish takes massive amounts of cultured algae. Hatcheries are the lynchpin in the shellfish aquaculture industry, and they require energy to raise water temperatures, keep seawater systems running, and provide bright lights for the algae as some of the requirements. Attention to detail is vital, and labor to keep everything clean is an important component – and expense.

Towns experimented with intertidal bottom boxes, covered with netting to protect the tiny seed from predators, and with floating sand-filled rafts. It was discovered that quahaugs need pressure against their shells, such as from sand, to thrive. They do not do well in floating bags or bottom trays that have no sediment around them. The floating rafts worked very well for both growth and survival but took a lot of space in areas where competition for space was keen. The intertidal boxes needed effective predator exclusion methods, but site selection was extremely important for success.

After a decade of field work, some towns created small hatcheries that operated on limited budgets. The most successful was the Martha's Vineyard Shellfish Groups' solar-assisted hatchery located on Lagoon Pond in Tisbury. The hatchery has been in continuous operation since

1978 and produces millions of quahaug, bay scallop and oyster seed that is distributed to five island towns according to Rick Karney, Director (emeritus) of the Martha's Vineyard Shellfish Group since 1976. In 2015, the MVSG distributed ten million seed quahaugs, eighteen million seed scallops and over a half million seed oysters to island communities.

**Martha's Vineyard Shellfish Group's municipal shellfish hatcheries. Their pilot hatchery, established in 1978, is on the left and their solar hatchery on the right has been in operation since 1980. (Photo by author)**

Several towns embraced the upweller nursery method when it appeared in the late 1980s. With upwellers, seawater is pumped into tanks. Containers called silos are designed with many shapes, sizes, and materials, but all have netting on the bottom that holds tiny seed shellfish and a hole near the top for a pipe serving as a water outlet. The water flows into the tanks, up through the silos and seed and out through an outflow pipe, back to the bay where it originated.

In this continuous flow-through system, the seed is constantly feeding on the natural plankton in the water. Sand is not

**Upweller – container, called a silo, has netting on the bottom that holds seed shellfish. Water is pumped into the tank and flows up through the silo and out to a trough and back to the bay. (Photo by author)**

required since the seed, crowded together in the silos, exert pressure on one another like the sand – a somewhat strange discovery, but it works. This space-saving system could handle smaller seed in greater numbers than the field gear. But there was a cost in electricity and personnel to maintain the system.

The combination of upwellers and field apparatus used successfully by the towns became the basis for commercial quahaug aquaculture, a highly valuable industry.

# Chapter 18
# Mussels Above and Clams Below

The commercial shell fisherman grunted acknowledgement as he ambled, bowlegged, across the sand in his rolled-down hip boots. As I walked toward him, streams of water occasionally spurted into the air from clams below the sand. Conversation was clipped as usual as he glanced at the groups of telltale siphon holes on the productive clam flat, an intertidal exposed sand bar in Nauset Harbor. He stopped and spread his legs, bent over like an offensive center on a football team, and started to dig.

With his clam hoe, a short-handled spading fork bent inward at about a seventy-degree angle, he dug down one and a half to two times deeper than the length of the tines and double the width of the fork. Acting like a human backhoe, he created a two-foot-wide trench in front of him as he worked his way toward the pencil-sized elliptical holes in the sand. With skill and care gained through years of experience, he placed the fork with its angled tines either in front of or behind a cluster of holes to avoid piercing the fragile brittle shells that would render the clams unmarketable. He plunked the legal-sized ones, two inches in longest diameter, now freed from their burrows, into his galvanized wire bushel basket.

The fisherman filled the trench as he excavated, leaving a mounded line of loosened sand and small clams behind him. Once a bushel was filled, he brought it to his anchored skiff, placed the wire basket next to the boat in water deep enough to cover the clams, and went back to fill another bushel basket.

The fisherman worked for several hours in this incredibly uncomfortable, backbreaking stance. It was easier to dig long trenches than to start new ones, so he continually worked his way around the flat, always looking at the siphon holes to dig in the direction of the greatest concentration. The sublegal-sized clams remained in the mounded sand and were fair game for the gulls milling around until the tide changed. Clams need water to dig themselves back into the sand. They cannot do it in air. When the tide turned and began to flood, the clams extended their foot, flipped upright, and disappeared beneath the softened sand.

Gulls were not the only predator seeking something to eat. Clams are prey to many predators – fish, crabs, starfish, and snails – and each has its own way of getting at its food. Moon snails were one of them on the flat that day in the 1980s.

Moon snails have a rather large foot with which they glide across the sand. One moved slowly and then stopped. A clam, not quite two inches long, was just outside the loosened sand mound on the hardened dry surface and could not dig back in without water. The moon snail glided over the clam, its foot enlarged to about three times its original size. Within ten minutes, this amorphous, beige, gelatinous mass had completely enveloped the clam, imprisoning it.

Moon snails possess a device known as a radula, an odd type of drill. It has been described as a minutely-toothed chitinous ribbon, the same substance as a lobster shell that, with acid, is used to bore a distinctive countersunk hole through the shell. Single clam shells with the telltale hole are a clear sign of moon snail predation. Starfish use a different method, employing powerful suction to pry the shells open to get at the meat within. Crabs crush the shells.

This twenty-acre flat at Nauset Harbor produced huge numbers of clams during some years but not every year. The location of "sets" of clams changed from year to year in the rapidly changeable harbor. As the Nauset Harbor barrier beach migrated farther north, clam production became spotty on the bar, and fishermen searched for new bars producing clams. Hunt and seek was the name of the game. Clams in the pure sand near the inlet grew fast, with extremely brittle shells, unlike clams in more silty or muddy sediment farther away, where they grew more slowly with correspondingly thicker shells. Allowed to harvest two bushels a day, each bushel containing about five hundred clams, the career commercial fishermen worked the bar every day they could. Some college kids clamming during the summer also worked the flats.

Softshell clams have historically been simply called "clams" on the Cape. They are also known as "steamer" or "piss" clams, for the streams of water that spurt up when people walk on top of them on the flats, or "whole belly fried clams" on menus. The regional historical moniker differentiates them from the hard-shelled quahaugs, but there are also a lot of other types of clams. For instance, clam strips are made from sea clams, a different large, mostly deep-water species. Razor clams are a sweet treat known more for their sharp shells than their taste. Mahogany quahaugs, a deep-water oceanic species that takes a long time to reach sexual maturity, are also processed for clam products.

Unlike quahaugs, softshell clams set gregariously – preferring to set with other clams in huge numbers. Unfortunately, not only do they set with others of their own kind, they set on top of one another. Small, illegal-sized clams are on top, and larger, legal-sized ones are below; this means that many smaller clams are destroyed in the process of digging to get to the legal ones.

Clams were dug on exposed tidal flats for generations. Shucked clams were salted and packed in barrels for the bait industry. In 1841, John Warner Barber, wrote in his book *Historical Collections Being a General Collection of Interesting Facts, Traditions, Biographical Sketches Anecdotes, etc., Relating to the History and Antiquities of Every Town in Massachusetts with Geographical Descriptions*:

"Between one and two hundred of the poorest of the inhabitants are employed in this business; and they receive ... three dollars a barrel for digging the clams, opening, salting them, and filling the casks. A man by this labor can earn seventy-five cents a day; and women and children are also engaged in it. A barrel of clams are worth six dollars; the employers, therefore, ... obtain a handsome profit. A thousand barrels of clams are equal in value to six thousand bushels of corn and are procured with no more labor and expense."

A barrel of shucked product equaled twelve to eighteen bushels of clams in the shell. Since the hand-digging harvest method has not changed, the thousand barrels collected in a year demonstrates not only incredible stamina by today's standards, but also indicates much greater clam productivity on the flats. Barber's description points to the economic circumstances under which people would resort to digging clams. It kept them going.

**Fishermen digging barrels of clams in intertidal flats. (Courtesy of Cape Cod National Seashore)**

There are no reliable statistics about how many people in the nineteenth century were engaged in clamming or about the amounts harvested. But when bushels are counted now instead of barrels, recent production seems to pale compared to a century earlier.

Mechanized clamming, first seen in the Chesapeake Bay, appeared in Cape waters during the last quarter of the twentieth century. Called hydraulic harvesting or pumping, local shell fishermen used gas engines to pump water. They fashioned a wand at the end of the outflow pipe to create a single powerful jet of water, similar to present-day power washers, or built a manifold with approximately eight jets to cover a greater area. Water jetted into the substrate loosened and liquefied it, liberating the clams from their burrows and onto the sediment surface. The men worked in teams. One walked backwards, liberating the clams with the apparatus, while the other walked forward, raking up the clams and dumping them into a plastic container. When it was nearly filled, they culled the clams onboard a boat, keeping the legal ones and tossing the smaller ones overboard. The new method made it possible for subtidal clams to be harvested from areas that had previously been inaccessible, a tremendous additional area. However, harvesting the subtidal clams that way removed valuable spawning stock for future sets.

Disagreements sprang up concerning the efficiency of the new method. A time-honored opinion held by most shell fishermen and managers was that "turning over the bottom" was conducive to shellfish setting, and that the "roughened" sediment allowed the tiny shellfish to anchor themselves to something that a smooth surface did not, explained David Belding. Pumping did just that.

Some shellfish officers believed that opinion was correct but that timing was crucial. Pumping was an appropriate harvesting method, but not during the spawning and setting time, especially from May through July. That effect was less pronounced at other times of the year.

It is probable that pumping blasts the tiny seed clams out of the sediment. They may land in heavy, black, lethal muck; or on the sediment surface where they are easy prey for small fish that congregate around a pumped area; or get smothered by the washed out sediment. In any case, they die.

However, the spawning time is also the season of greatest market demand and highest prices. If repeated pumping at that time of year reduces the number of seed clams in order to harvest the adults at the highest price, the population dwindles. At other times of the year, the clams are usually big enough to survive. If, however, bait fish are active, they often take a bite out of a clam's foot as it digs back in. Once injured, the clam can't dig and stays on the surface, easy prey for crabs.

Once again, increasing harvest efficiency through a technological advance decreases overall productivity if done at the wrong time.

～～～

Massachusetts towns have experimented with various methods of clam propagation for decades, going back to Belding's work in the early 1900s. Cape communities used netting to prevent predation for some experiments in the last quarter of the twentieth century, but netting was also used to assist the gregarious setting concept.

Microscopic clam larvae floating in the currents go through a metamorphosis. They lose their swimming organ, gain a foot and a hard shell, drop to the bottom, and search for a new home – the process of "setting." At the Orleans Shellfish Department, we wondered if the miniscule clams detected a chemical cue in the water that adult clams are nearby, would they set under the netting with other clams, protected against predators? The answer to that was yes. How about using the netting in areas where clams were not in the general vicinity but were a mile or so away? The assumption was that larvae go where the currents take them, so there should be some in the water. It didn't work. Adults had to be there first, covered with netting; then the baby clams would set. It became obvious, though, that covering large flats with netting was impractical and cost-prohibitive. So researchers and towns turned to Belding's suggestion – turning over the bottom to encourage setting.

Albert Redfield, a Woods Hole Oceanographic Institution researcher, hand-plowed plots in Barnstable Harbor in 1960, as illustrated in the WHOI magazine *Oceanus*. And Eastham's shellfish constable Phil Schwind plowed there in the early 1970s. A decade later, we mechanically plowed large areas in

Dr. Albert Redfield using a hand plow for clam experiments in Barnstable Harbor. (Courtesy of Woods Hole Oceanographic Institution)

Orleans. In late summer, we transplanted small clams, collected from areas of high natural set, in the softened sediment on an incoming tide so the clams could dig themselves in before birds could feed on them.

**Plowed area for transplanting seed clams without netting. Clams planted on incoming tide dug in quickly with high survival, especially if they were in the furrows. (Photo by author)**

Belding included a photo of a flat in Rowley, north of Boston, with fifteen hundred clams per square foot when they set. That amount later diminished to four hundred, of which only five percent reached maturity. When clams occasionally set in that density, they provide an opportunity for communities to thin them, similar to garden thinning, since such a small percentage will reach maturity where they set. Then the town can transplant some of them to other areas and either protect the seed clams with netting in small plots or plant them in larger areas at low density. Plowing, loosening the substrate, worked well. When seed was transplanted on an incoming tide after plowing, clams dug in quickly and survival was generally good without the expense of netting.

Massachusetts communities learned much about clam propagation and protection from researchers in Maine, where clams are the primary shellfish species. The Maine Department of Natural Resources sponsored clam conferences, usually in the Boothbay region, in the 1970s and '80s. All of the presentations were geared in some capacity to clams. The University of Maine's Darling Marine Center actively researched ways to increase clam harvests, and control predators such as green crabs, as did the state's Department of Marine Resources. In 1987, the Beals Island Regional Shellfish Hatchery opened as the first public shellfish hatchery in Maine, a prominent research facility spearheaded by Dr. Brian Beal from the University of Maine at Machias. The modest hatchery has grown tremendously, becoming an incorporated nonprofit in 1997 and changing its name to the Downeast Institute for Applied Marine Research and Education to better reflect its mission.

Clam seed propagation came to Salem State University in Massachusetts through research conducted by Dr. Joseph Buttner and his students as a direct result of the work in Maine. While not as prevalent as for quahaugs and oysters, clam culture continues in Massachusetts on a limited scale.

Clams and mussels often share the same space above and below the sediment surface. But mussels are a different story.

~~~~~

The winter of 1976-77 was harsh. All of the bays were frozen for fourteen consecutive weeks. The only open water in Orleans was the fast-moving, thirty-foot-wide, three-foot-deep creek between Mill Pond and Robert's Cove, a part of the Nauset Harbor complex. The rock foundation of an old dam blocked the flow between the cove and the pond except for two openings. Those openings provided gushing water with the tide, and the powerful constricted flow prevented a two-acre segment of the small channel on the Robert's Cove side from freezing. Mussels were not a popular shellfish, but with nothing else available, fishermen were forced to harvest them from the channel for any income that winter.

About a dozen men used pitchforks to break up large clumps of mussels in the channel. They tossed the smaller clumps into a small skiff, and when it was filled, they dragged the skiff to shore to bag the mussels. With no marketing standards at the time, the men did only a cursory cleaning, dubbed the "Orleans stomp." They put the mussels in wire bushel baskets until they were half full, and with the baskets covered with water, they broke the clumps apart with the heels of their boots. Seed mussels, pebbles, snails, and other debris and mud fell out of the basket, leaving mostly mussels.

As soon as the bays thawed, the men moved on to more lucrative fisheries. That summer, however, a pair of fishermen used a scallop drag in the channel. They shoveled everything that came up in the drag – rocks, mud, snails, broken shells, unbroken clumps of large and seed mussels – into bags to sell for about $1.25 per bushel. By taking everything – seed mussels and the substrate they needed to set – the two men emptied the channel of mussels. The mussels never came back.

Mussels appeared elsewhere in Nauset Harbor, though. As local chefs added mussels to their menus, prompted by popular ethnic cuisine in city restaurants, the demand rose. The price per bushel rose. The number of people involved in the fishery rose. As one of the few towns with high-quality natural mussels, and in an effort to maintain a fishery, the Orleans Shellfish Department suggested regulations: a minimum size, a harvest limit, and a requirement to sort, cull, and return seed to the water on site. The suggestion was met with animosity. It was an unregulated fishery on the state and town levels, and the fishermen wanted to keep it that way. The shellfish department capitulated while reminding everyone involved that the regulations were meant to preserve the fishery for the future.

By the mid-1980s, demand escalated and the wholesale price increased from ten dollars to eighteen dollars to over twenty dollars per bushel. As more people entered the fishery and the number of mussels predictably declined, the shell fishermen themselves requested regulations, suggesting the same ones we had proposed several years earlier. They were quickly enacted.

Since then, mussel production has waxed and waned from natural population fluctuations and for other, unknown reasons. It often depends on the number of green crabs and of eider ducks. Mussels are the favored food of the diving eiders that winter over in Nauset Harbor. High tide at Nauset is no real challenge for them, and they can wipe out a huge bed in a single winter. The price paid to fishermen has waxed and waned too; the price quoted in 2016 was eight to nine dollars per bushel.

Mussels are another shellfish species that set gregariously and are incredibly prolific. They withstand waves and strong currents, but they need something to attach to. Once a few are anchored to rocks or pilings with their strong byssal threads, others set on top of them, forming large clumps. Intertidally, they create a reef or "bed." Intertidal mussels grow more slowly than subtidal ones and tend to contain irregularly shaped pearls which are a detriment as a seafood to the consumer and a bonus to dentists.

Mussels and clams can sometimes produce tremendous sets primarily because of their gregarious setting traits. While clams and mussels can be extremely productive, they also have periods of inexplicably poor sets. Clams did not set appreciably during the 1940s. During that time, the war years, Orleans Shellfish Warden Elmer Darling urged residents to harvest and eat the prolific but detested mussels as an abundant source of protein. As soon as the war was over, Darling urged residents to get rid of the mussels in order to enhance areas for clams to set once again.

In contrast to quahaugs, clams and mussels have shorter lifespans. Thus, they have more variable populations. Natural fluctuations – environmental conditions, predator and prey dynamics, harvest, and other factors – play a part in whether these shellfish will set and live to be harvested.

Harvesting wild mussels has declined tremendously. The vast majority of mussels sold now are cultured, primarily in Maine and Canada, although a large mussel company operates in Massachusetts waters harvesting both wild and cultured mussels.

By the mid-1980s, green filamentous seaweed was covering the flats and mussel beds and smothering the shellfish. At first, we thought it was because the inlet was migrating so far north. But we were seeing the seaweed throughout the estuary and were concerned about the cause, which was not identified or hypothesized until late in the decade. The green seaweed indicted that nutrients, primarily nitrogen, were increasing and fertilizing the green plants. It was not a good sign, and it turned out to be one of many ominous signs.

Chapter 19

# Nuggets of Gold

Trucks towing boats on trailers formed a steady line to Rock Harbor on September 30, 1975, the day before the scallop season opened in Cape Cod Bay. They came from landings on Pleasant Bay and Town Cove to launch their boats at Rock Harbor. A large number of scallops were in a guzzle, a relatively small section of deep water on the town line between Orleans and Brewster. It was a rare occurrence for the two towns.

Scallop fever had hit three adjacent towns: Brewster, Orleans, and Eastham. Orleans and Eastham share waters and have a joint fishery. Residents of those towns can fish commercially or recreationally in both towns with a proper license from that town at the resident fee. People who are not residents of either town have to pay a higher fee for a recreational license and cannot fish commercially at all.

Bay scallops live for only two years. They are born in the summer, grow until the water cools in the fall, put on a visible growth ring when they start feeding again in the spring, and then spawn in the summer. They are harvested in the fall after spawning and are dead by the following spring. State law sets the scallop season as October 1through April 30 but towns can adjust the opening date. Some towns have delayed opening dates in different bodies of water within the same town to spread the season over a longer time, prevent a glut on the market and to allow the scallop meats to increase in size. Scallops are one species for which over-harvesting is not possible if they have the visible raised annual growth ring. There were so many adult legal scallops in the bay with that growth ring that it was unlikely they all would be harvested before December, when days on the water would be very limited. And they would be dead by spring. The shellfish departments wanted to allow as much time as possible to harvest the scallops, so we opened the season as early as legally possible, on October 1, but we allowed no more than two permit-holders per boat.

Brewster residents could not fish commercially in Orleans, but on the water, the town line was pretty difficult to enforce. To get around that issue, many boats carried one commercial license-holder from Brewster and one from Orleans or Eastham. All of the boats had to land their catches at Rock Harbor, so from a practical point of view, it didn't really matter where the scallops came from, and it was an economic boon for the whole region.

The scallops were only a mile or so from land, but if the boats stayed out too long, they could not get back into the harbor over the exposed tidal flats at low tide, so harvesting was done during the four to five hours of high tide. Small boats, in the sixteen to twenty-five-foot range, made up the majority of the fleet, although a small number were thirty-five to forty-five feet. Each permit-holder was allowed ten bushels of scallops. Twenty bushels left the smaller boats with precious little freeboard, a dangerous situation. Cape Cod Bay in the fall was already unsafe for these boats, which were more appropriate for sheltered bays in the summer. Bad weather can come up quickly in the bay with its vast open water all the way to Boston. Boats being swamped was a real possibility. Miraculously, it never happened.

The weather changed abruptly one day. The skies turned an ominous steely gray, and the wind came up fast. No one wanted to be called "chicken" as the first one to turn tail and head for the harbor. But the boats settled lower with each bag of scallops aboard, and the waves chopped higher as the weather deteriorated. Finally, one small boat turned and headed in – followed by a long line behind it.

Even ten bushels per person per day weren't enough for some of the fishermen, especially those with larger vessels. Within a couple of weeks, they started using grain bags, stuffing as many as three bushels into each bag and hiding them, or hiding extra bushel bags in the boat to be retrieved at night, while showing us regulation bushels. The men who were playing by the rules complained that other fishermen were using big bags or taking more than the limit. After proving the validity of the complaints, we put up signs that stated: "Bushels Only" beginning the following day.

Scalloping lasted until mid-December when the weather changed, the temperature dropped, and the bay was close to freezing over. The town removed the harbor floats for the winter. It was all over. And it never happened again.

**Bushel of scallops with growth ring evident on the shells. (Photo by author)**

The buzz started shortly after Labor Day. Men sitting at the counters of the local coffee shops in the early morning or at the bars later in the day whispered conspiratorially to one another.

"I threw a drag over today."

"Oh yeah?"

"There's a few out there. Keep it to yourself."

" 'Bout time. We haven't had a good season in ages."

Scallop season was fast approaching. Every year, soon after most of the summer tourists had left Cape Cod, fishermen began thinking about scallops. Illegal to harvest until later in the fall, they nonetheless took a chance during early mornings or on foggy days, when shellfish wardens were unlikely to see them, to do a "test drag," looking for the succulent bivalves. If they found some legal scallops, they tried to keep it to themselves. But word eventually got out, and before long "scallop fever" – the anticipation of a good scallop season – gripped the community.

People checked their gear, made repairs, and got ready. Then, on November 1, opening day on Pleasant Bay, they thronged to every landing and headed out at dawn as if someone had fired a gun to start the race.

It looked like the fall of 1979 would be a pretty good scallop season. And it was – for about six weeks. It was also an odd year. The bay was loaded with scallops, most of them enormous. But they did not have a "raised annual growth ring," the visual indication on their shells that indicted they had gone through their one and only spawning season. Scallops that did not have that ring on were considered immature "seed" and were illegal to harvest. Unfortunately, about ten percent of them had a ring but it was tiny, very close to the hinge, that we called "ring at the hinge" scallops. Fishermen argued that those were legal. Those odd scallops just added another layer of complexity to an already difficult situation.

Fishermen lamented that the large seed would never live through the winter to spawn the next summer. They said it had happened before and that the scallops never made it to the next year. Surely "they should be taken" was the sentiment expressed time and again at the landings where we checked the catches. Fishermen also acknowledged that they had taken really big scallops like that in the past, and shellfish constables had always "turned a blind eye." The shellfish constable said that wasn't going to happen this time and repeated that they were illegal; size made no difference. He gave the fishermen a day of warning and said that any illegal scallops would be confiscated after that. The fishermen wanted proof that they were seed. When I pointed to a black pigment covering the scallops' gonads, a clear indication that they had not spawned, that was not good enough. The fishermen demanded more proof – anything to delay the inevitable action of shutting them down for the winter and denying them one of the few fishing options open to them at that time of year.

The Orleans Shellfish Department was left with no choice. Following several boisterous and highly-charged meetings, the shellfish constable recommended that the selectmen close the bay to harvest and protect the remaining seed, which they did. Later, we did prove, through microscopic analyses of the gonads performed by Dr. M. Patricia Morse and her team at Northeastern University, that the scallops, including the ones with the ring at the hinge, had not spawned and were seed. During the next season, fishermen harvested a record 4,520 bushels worth $190,000, and similar harvests occurred in 1981 and 1982.

The shellfish department believed that closing the bay and protecting the seed for potential future spawning and harvests was an appropriate management decision. Then, in 1983, Pleasant Bay produced a scallop bonanza of 37,500 bushels at a wholesale value of $1,550,000 with smiles all around. The number of new trucks in town was a visible symbol of the economic benefit. It never happened again for that size harvest or for even a reasonably-sized one from Pleasant Bay.

**Scalloping in Pleasant Bay close to shore. (Photo by author)**

Saturday dawned bright and clear on the Columbus Day weekend in October 1985. Low tide was late morning. It was a picture-perfect day for the traditional opening of scallop season at Town Cove. By eight o'clock, the crisp fall air had begun to warm up. Asa's Landing was packed with cars, and Gibson Road leading to the landing was filling fast. A couple of hours later, cars lined the road for nearly a half mile. Word had spread. There were lots of scallops in the cove for a change. They were almost always smaller than those in Pleasant Bay, but nobody really cared. Later, when the people got home and had to open them all, it might be a different story. But for scallops worth about fifteen dollars a pound in the market at that time and with about six pounds to the bushel, the five-dollar resident shellfish license was a bargain, well worth the effort to get and shuck scallops for the freezer.

Scalloping in Town Cove was mostly geared toward family permit-holders who were allowed a bushel a week. The flats and shallow areas were reserved for them. Commercial permit-holders were allowed to fish in deeper water where dragging was appropriate.

**Two recreational fishermen sharing the weight of a bushel of scallops. In the background, many residents share the bounty. (Photo by author)**

Hundreds of people walked around in the shallow water, picking up "nuggets of gold" with quahaug rakes or ordinary garden rakes. Clad in jackets or sweaters, and hip boots or waders – or old sneakers for the really hardy souls – they searched for areas not already stomped on by other fishermen to get their one-bushel limits. Many took half their limits – a bushel got pretty heavy while walking back to the car. They knew they could come back the next day for the second half.

The smiles were infectious as I stood at the landing checking licenses and talking to the people coming back with their scallops. It was a magical weekend. It was the largest recorded harvest there. And it was the next-to-last time for such a miraculous fall bounty in the Cove.

Cape Cod Bay in 1975; Pleasant Bay, the largest bonanza of them all, in 1983; and Town Cove in 1985. All of them had scallosps in the late 1970s or mid-1980s – and then nothing. What was going on? It turns out Orleans was not alone in a decline of scallops after the mid-1980s.

Although scallop populations commonly fluctuate like yo-yos, the latest manifestation is that the low end has lasted longer than at any previous time. Part of the reason for scallops' value in Massachusetts is that they are not always available, a function of supply and demand. Belding has described Cape Cod Bay as the northern boundary for the species. The scarcity of scallops throughout southeastern Massachusetts prompted the Barnstable County Cooperative Extension to investigate the downturn in 1999.

Surveys were sent to managers and researchers in the region representing twenty-two towns, the Massachusetts Division of Marine Fisheries, Martha's Vineyard Shellfish Group, and Westport Water Works to gain some insight. Nearly half of the respondents had more than ten years of experience, and an additional one-third had over five years of experience. One-half to three-quarters reported observing changes in environmental factors: a decrease in eelgrass; an increase in epiphytes, seaweed growing on eelgrass leaves; an increase in seaweeds; increased sediment changes from hard sand, shells, and silt to soft mud; and an increase in plankton blooms. All of those observations represented shifts in habitat associated with increased nutrients in estuaries. Research in Pleasant Bay and Waquoit Bay and the 1991 Buzzards Bay National Estuary Project all pointed to water quality degradation resulting from coastal development.

In light of the responses, a second phase of the project involved developing a restoration plan for Barnstable County. Shellfish officers identified potentially suitable areas where deploying spat catchers, onion bags filled with gillnets for scallops to set on, might be successful. Spat catchers are used to concentrate larvae in a protected environment where they can set and grow to small seed and planting seed. There were three bottlenecks for implementation: lack of hatchery seed production for planting seed; cost; and impracticality of capturing natural seed because of the low numbers of spawning adults in the region. Without spawning stock, a restoration program would not work.

Martha's Vineyard, on the other hand, has been a scallop success story. Martha's Vineyard Shellfish Group's publicly-supported hatchery, is the largest producer of its kind in Massachusetts. The hatchery has successfully produced millions of scallop seed each year for the Vineyard's towns as well as quahaugs and oysters. They also experiment with other species and conduct research on many issues. The yearly addition of millions of seed scallops to Vineyard waters through annual plantings from hatchery-raised and town-managed nursery systems means that the Vineyard enjoys a more stable scallop population.

Nantucket has one of the few remaining, though dwindling, natural scallop populations. Nantucket has followed the Martha's Vineyard lead and has developed a robust scallop propagation program. They have also experienced times when the scallops had a raised growth ring, but it was tiny, located very close to the hinge just like the ones in Pleasant Bay. Biologically, these scallops were seed, often very large, that had not spawned, even though they had a legal raised ring. When they occurred, the temptation to take them and not wait another year was huge, causing a substantial enforcement issue. Dr. Valerie Hall completed her doctoral dissertation in 2014 based on the reproductive cycle of bay scallops in Nantucket with special emphasis on the "nubs" as these odd scallops were dubbed there. Her research corroborated that of the Orleans experience especially with respect to the reproduction potential by leaving the scallops to complete their natural spawning activity.

The combination of the two Barnstable County projects served to identify scallops as an indicator species that is very sensitive to environmental changes. They also indicated a relationship between scallops and eelgrass. Scallops set and attach to eelgrass blades as tiny seed and remain there until they are too heavy, when they drop off. Over time, if the eelgrass cover decreases, the scallops decrease, even though they may attach to other seaweeds as well.

When water quality is degraded, a chain reaction takes place. Epiphytic seaweeds, small plant species that grow on other plants, attach to the eelgrass blades; other seaweeds multiply and grow among the grass within an eelgrass meadow; and thick phytoplankton clouds the water. These three degrading factors diminish light penetration to the eelgrass. Thus, the eelgrass dies, and the scallops can't set.

We were beginning to understand the relationship between human activities on the land and the effects in the water, and it was not a positive relationship. We were adding too much fertilizer in the form of nitrogen from our septic systems, our storm-water drainage, and our green lawns. Not only was the increased marine growth a problem, but when those seaweeds died, they consumed oxygen, depleting the total amount in the bays. Virtually all of the estuaries in coastal Massachusetts were seriously out of whack, and only a concerted effort could save them.

## Chapter 20
# Farming the Waters

The September day in 2013 was beautiful with clear sunny skies, little wind, no humidity and temperature in the seventies. Orleans' annual "Celebrate Our Waters" festival was unfolding, with more than fifty free weekend activities set for land and water venues all over town.

One of the activities was a tour of an oyster grant on the flats of Cape Cod Bay. Wendy and Kyle Farrell, a couple who leased a small piece of sandy bottom, had been in the shellfish aquaculture business for just a few years. As Wendy patiently answered questions from the fifty or so attentive people who had walked a half mile on the flats to reach the site, I inspected the various types of gear. Some of it was for nursery culture to grow the oysters from very small seed to harvest size, and some of the equipment was intended to attract a natural set.

There were "oyster condos," vinyl-coated, wire-mesh "cabinets" constructed with six shelves, each holding a plastic mesh bag of small oysters. Another type of nursery equipment was hard,

**Wendy Farrell pointing out oyster nursery gear: covered plastic trays and oyster condos. (Photo by author)**

plastic square trays that could be covered or not, with one-inch holes to allow a free flow of water, designed to hold larger oysters. "Chinese hats" were used to catch wild oyster spat, tiny oysters that attach to something hard like other shells or cement. These round plastic disks were arranged vertically and held on a plastic pipe with a couple of inches between each disk, each covered with a thin veneer of cement. Wagons with large tires were on hand to transport the gear and harvested oysters to shore. It was clear that the Farrells were experimenting to see what would work best for their situation.

**Chinese hats to catch a natural set of oysters.
(Photo by author)**

The Farrell grant is one of five shellfish grants, or leases, on the unproductive flats that had been set aside for this purpose in a 1980s shellfish management plan. However, the grants had not been issued until well over two decades later. By then, oyster culture had gained in popularity along Cape Cod Bay shores as a potentially profitable enterprise despite inherent challenges with this particular site. Now, all five grants have been leased, and each farmer is allowed to expand to two acres. Large, bright orange buoys mark the corners of each grant, the only visible markers at high tide. When the ten-foot tides recede, rows of gear cover the individual grants.

It is not an easy life. The farmers cannot access their grant by vehicle during daily operations but must lug everything out by hand or bring it out by boat and work the tide. They must pay close attention to the weather and remove all the gear before the bay freezes over, or they risk losing it all. And, because of the visibility and public accessibility, they must answer endless questions posed by people walking the flats while they do their work during the summer. The price they receive for their product does not compensate them for the free information they impart. To Wendy and Kyle, it is worth it. Wendy clearly enjoyed talking to the people who visited the farm that September afternoon as she pointed with obvious pride to the crop they were nurturing.

Applications for small shellfish grants in Pleasant Bay were coming in one after another during the 1980s. Many of the applicants were new residents who had not been on the Cape for very long and knew precious little about growing shellfish, but they owned waterfront property. At the same time, other waterfront property owners were applying for permits to build private docks in public waters. When questioned, it turned out that what these owners really wanted was to keep the water in front of their houses for their exclusive use.

An ancient state law, the Colonial Ordinance of 1641, stated that ownership of any waterfront property extends to the mean low water mark, but the ownership property rights did not include the water or the resources, such as shellfish, in the strip between high and low water. That was within state jurisdiction and the state ceded management of those resources to the towns. The public had specific rights of access for "fishing, fowling, and navigation" between mean high and mean low tide. Anyone could fish, harvest shellfish, shoot waterfowl, or operate a boat anywhere along that strip of shore.

The property owners didn't like that stipulation. They didn't want to see or hear people shell fishing. They didn't want to see gear in the water. They didn't want their tranquility and enjoyment marred in any way. They were paying high taxes to live there, and they felt entitled. It was a classic NIMBY attitude, "Not In My Back Yard."

Shellfish often occupy the intertidal space between the edge of a marsh and the low-tide line. Docks impede navigation, and their construction and continual presence and use, especially by motor boats, destroys shellfish habitat. The presence of shellfish is grounds for local conservation commissions to deny dock applications because of the Massachusetts Wetlands Protection Act. Shellfish grants receive even greater scrutiny because the sites must also be approved by the state's Division of Marine Fisheries. None of those applicants for shoreline shellfish grants during the eighties satisfied both town and state regulations, and their applications were denied. At the same time, a few applications were submitted by people who wanted to get into aquaculture because wild harvests were at a very low ebb. Finding suitable sites was difficult, and those applications were also denied because they did not meet the state's standards for site selection.

Any area designated for shellfish aquaculture had to meet both state and local criteria. First and foremost, state regulations dictated that only areas that "would not cause adverse impacts to the town's shellfish resources" could be leased as shellfish grants. This meant that nonproductive areas were the only ones eligible for leasing. State biologists surveyed proposed lease sites and counted any of the five major commercial species they found — clams, quahaugs, oysters, scallops, and mussels — to determine productivity. The grant also had to be away from the immediate shoreline where the public had rights of access through the Colonial Ordinance. Orleans required the site to be outside prime boating areas, away from channels, and essentially in an "out-of-the-way" place. Lessees required the landowners' permission to access the sites by land. Other towns set their own restrictions.

While motoring around the bays in the Orleans town workboat, I thought about the increase in grant applications and the efforts that went into the process to approve the sites, most of which ended in denials. The process wasn't working. Perhaps a suitable solution would be to find an area that could be set aside for several grants adjacent to one another and permitted for commercial aquaculture. I suggested two barren areas in Pleasant Bay and one in Cape Cod Bay. All those areas had long histories of poor to nonexistent production for any of the five major species.

There's always a reason that an area is barren — sediment, current, and circulation patterns; something preventing natural setting; predation; or a combination of factors. The theory was that if an individual wanted to invest his own resources for growing shellfish in a less-than-ideal situation and was successful, the shellfish he grew would become parent stock before being harvested and potentially repopulate public areas surrounding the grant. Both the individual and the public would benefit.

Commercial fishermen who attended the Shellfish Advisory Committee meetings fought vociferously against the plan on the grounds that it would take territory out of the public domain and put it in private hands and those areas would never again be open to the public. They thought the proposed plan was heresy. We were suggesting that public waters be leased to private individuals for exclusive rights to grow shellfish for profit. Yet, the areas we were suggesting were barren and had been for generations. There was no shellfish to harvest.

After two years, the selectmen approved an aquaculture plan that set aside two twenty-acre sites in Pleasant Bay and one ten-acre area in Cape Cod Bay for shellfish grants, each to have a maximum of two acres per individual lease. No areas were included for Nauset because there were few barren areas that were large enough.

All of the grant sites had issues besides non-productivity. One in Pleasant Bay was protected from bad weather but was poor for shellfish growth. Quahaug seed planted there took five to seven years to reach the legal size of two inches in longest diameter, more than twice as much time as in other locations. The second area was far enough away from any type of security so it could not be protected against poaching. The third site in Cape Cod Bay was on the flats, vulnerable to harsh wave action, winter ice, plus vandalism and poaching because it was near an easily accessible public area.

These inherent problems with these sites meant that potential farmers were slow to take the challenge. But a few enterprising young men decided to give it a go. They started with quahaugs.

Cape and Islands municipal propagation experiments of growing hatchery seed with both bottom culture — seed planted in the sediment and covered with netting — and off-bottom culture — some sort of floating gear like the sand-filled rafts — showed promise. At the same time, wild-capture shellfish fisheries were declining everywhere. The Aquacultural Research Corporation, a local producer of the hatchery seed in Dennis, took note of the municipal efforts and successes. They partnered with individuals, many in Wellfleet, to grow quahaug seed on private leases. Growers planted the tiny quahaugs in wide swaths on intertidal flats and covered them with netting. It was the sunrise for a new aquaculture industry.

All was going well. Private aquaculture was growing by the late eighties and early nineties. Provincetown had a vast area that was unproductive and established large lease sites. It was a way to keep fishermen on the water and provide steady employment in a community where fishing had declined rapidly and where most other jobs were seasonal. Wellfleet increased its leases. Orleans, Eastham, and Barnstable established aquaculture development areas. As predicted, quahaug growth was slow in Pleasant Bay, several years slower than in Wellfleet where it took only three years to reach harvest age. That discouraged greater participation in the program in Orleans. But those involved kept at it and eventually realized a cash crop.

**Early shellfish grant sites under cultivation for quahaugs in Pleasant Bay before the growers switched to oysters. (Photo by Kevin Flynn)**

Then in 1995 a disease called QPX, Quahaug Parasite Unknown, decimated the crops in Provincetown. It was next discovered in Orleans in 2000, and finally a massive outbreak was detected in Wellfleet in 2004. In Wellfleet, a huge number of quahaugs were destroyed on purpose to prevent the disease from spreading. Painful as that decision was, it saved the burgeoning industry.

Shortly after those incidents, many quahaug farmers switched to oysters, and the industry mushroomed. As of 2015, the Division of Marine

Fisheries recorded nearly 350 growers, most of whom were cultivating oysters. The Cape Cod Bay region, including Cape towns from Sandwich to Provincetown and off-Cape towns of Plymouth and Duxbury on the west side of Cape Cod Bay, accounted for nearly 60 percent of the total acreage under cultivation statewide. Nearly every community on Cape Cod Bay has private shellfish grants, and grants are also located in Nantucket Sound and Buzzards Bay. The economic boon is tremendous, with the entire Massachusetts aquaculture sector contributing more than $23 million. Shellfish aquaculture provides jobs, food, and what are collectively known as ecological services, or environmental benefits, that are helpful to us all. They include filtering the water which enhances clarity and promotes eelgrass survival, removing nitrogen, and providing habitat for many juvenile fish and crustaceans.

Surveying and setting aside specific areas for shellfish aquaculture, now called aquaculture development areas, became common practice among Cape towns. The practice streamlined the often onerous permitting process and fostered encouragement for new farmers because the towns initiated the first step, site selection, although obtaining individual permits is still necessary. The result has been increased production and innovation in gear and techniques. Shellfish aquaculture also transformed many commercial fishermen into farmers, allowing them to stay on the water and earn a living while taking some pressure off the public resource.

Truro and Provincetown embarked on establishing an aquaculture development area of fifty subtidal acres straddling the town line, assisted by a team led by Owen Nichols from the Center for Coastal Studies in Provincetown. In this case, all permits would be obtained by the towns, and the individuals who lease sites would have to comply with certain operational requirements, but they would not have to go through the permitting process themselves. Growers must sink the gear during the time of year when right whales are in the bay since the development area is in potential right whale habitat.

Shellfish aquaculture faces unique challenges. Farmers have to contend with Mother Nature vagaries of weather, unwanted plants and animals, (biofouling) covering gear; predators, and competitors for food, to name a few. They also operate their businesses in leased public areas. Unlike many land farmers, who own their farms outright, everything done by a grower who leases public land is subject to public scrutiny, during the permitting phase before they obtain their lease and then every day as they operate their business. Land farmers are not subject to such intense scrutiny unless their farming practices impact people or resources off their own property – runoffs into streams or groundwater, for example.

Ironically, shellfish farmers require clean water to operate their business. They are keenly aware of the connections between the land and the sea because their livelihood depends on maintaining good water quality. Bacterial contamination can shut them down. Natural events like the red tide can shut them down. Human illness traced back to them or the bay they work in can shut them down. Despite the challenges, they work tirelessly in one of the most highly-regulated and green industries in the country to provide safe shellfish for consumers to enjoy.

One particularly vexing problem facing growers is naturally occurring bacteria, not associated with pollution, known as Vibrios, which thrive in warm waters. Two species are associated with shellfish-related illness: Vibrio vulnificus (Vv) and Vibrio parahaemolyticus (Vp). Illness is always more severe for people with compromised immune systems, who should not eat any raw proteins, especially raw shellfish during the summer. V. vulnificus is especially dangerous for those individuals. Thankfully Vv is very rare. Only about one hundred cases are reported each year, primarily in the Gulf states. Vibrios are among the fastest growing bacteria on earth and will double in less than an hour at ninety degrees Fahrenheit. Vibrio growth can be arrested by chilling shellfish after harvest, keeping them below fifty degrees. Each state is required to develop a Vibrio Control Plan that dictates how and when growers and harvesters must refrigerate shellfish. The plans dictate how soon harvesters must get their shellfish on ice or into refrigeration during the summer months. If an illness is associated with shellfish consumption, health officials trace it to the source. If several cases are traced to the same harvest area, it will be closed for harvest and shellfish from that area will be recalled and destroyed. Harvesters are the first link in the chain and must adhere to regulations that are designed to safeguard public health. However, everyone in the chain, including the consumer, is responsible for maintaining proper temperature standards. The industry recognizes that illnesses will shatter consumer confidence and destroy the markets that they have worked hard to establish. It is impossible to make raw foods risk free, but regulators and the industry have developed an effective cooperative program that minimizes the risk so consumers can enjoy raw shellfish with confidence. Consumers share the responsibility for shellfish safety and must remember to keep them chilled, especially in the summer.

Shellfish are grown in estuaries where there is competition for the same water: commercial and recreational fishing; boating of all types; and beaches, marshes, and rock ledges to explore. Our coasts are highly desirable. In 2010, 39 percent of the U.S. population lived in coastal counties, while over half the population lived in coastal watershed counties, where at least 15 percent of the land area has a watershed-based impact on coastal and ocean resources. Moreover, a whopping 97 percent of all seasonal housing in the U.S. is located either in coastal counties or the watersheds to the coast. Most of those counties are within fifty miles of the coast. People who vacation at the shore often come from inland areas, although exact numbers are elusive. Nevertheless, that's a lot of people with varying desires about enjoying the nation's watery playgrounds, and that influx of people and their impact is acutely felt on Cape Cod and similar tourist areas, where the population triples or quadruples during the summer season. What were traditional summer cottages during the early and mid-twentieth century changed to summer houses that were later expanded to larger houses, and some became year-round homes as vacationers decided to retire there.

People want shellfish in the market or at the restaurant, but they are generally unaware of what it takes to produce them. If growing shellfish in an area means potentially decreasing their pleasure, they often voice concerns. With all of the competing uses, estuaries are long on the number of people who enjoy them and short on adequate space for industries such as shellfish aquaculture to flourish.

Bottom gear is one thing. Floating gear is quite another. Oysters live on the bottom in their natural habitat, but single-oyster cultures suitable for raw-bar consumption require many different configurations of containers to keep the oysters off the bottom.

**Floating nursery gear on oyster lease. (Photo by author)**

Surface gear has many advantages, including no dependency on the tides, less predation by species that live on the bottom with oysters, and fast growth. One detriment is fouling, a serious problem with floating gear. Another is constant visibility. Lines of buoys and black plastic bags filled with oysters occupying many acres in a publicly-traveled bay can cause problems – from navigation to NIMBY. Use of the entire water column, not just the bay bottom but the bottom to the surface, raises the level of concern for others who use the water.

To counteract negative perceptions about shellfish aquaculture, the East Coast Shellfish Growers Association developed best management practices during workshops with growers, regulators, extension agents, and other stakeholders in all of the East Coast states. The end result was a guidebook describing how growers could operate their farms in an environmentally and socially-responsible manner. It begins with the all-important site selection and permitting process which includes biological and multiple-use social considerations. Some of the actions were matters of common sense, such as cleaning gear by air-drying or with salt brine to remove biofouling rather than using chemicals to kill the unwanted marine growth. Others were more subtle, such as being a good neighbor by not operating machinery like pumps or mechanical harvesters or shellfish tumblers early in the morning when sound is amplified over the still water. Farmers were encouraged to remove damaged gear from the water and dispose of it properly, to search for gear that had been driven offsite by storms, and to develop a basic "Good Neighbor" policy. The guide also provided a template for growers to create an individual farm management plan that enabled them to incorporate their best management practices into their business plans or operation policies, demonstrating to regulators, lenders, insurance adjusters, or skeptical neighbors that farmers were operating their farms responsibly.

Despite the problems, shellfish aquaculture is an expanding industry. South Carolina, Georgia, Florida, Alabama, Mississippi, and Louisiana embarked on programs to encourage new oyster farmers to grow single oysters for the half-shell market in areas where clumps and reefs of natural oysters are common. Delaware enacted its first leasing program and Maryland changed their laws to permit greater leasing potential while the Virginia industry has exploded during the last decade or so. That state now has five hundred lease applications pending for quahaugs and oysters. Florida is experimenting with marketing a new species – sunray venus, a type of steamer clam with a shell that, when cooked, turns a beautiful striped coloration, resembling a sunray through clouds. And New England aquaculture continues to expand. The East Coast Shellfish Growers Association tracks the industry, a daunting task to keep pace with the number of farms, number of employees, species and amount cultured, amount of acreage under cultivation and value. East and Gulf coast states raise American oysters. Aquaculture on the East Coast is basically a "mom and pop" operation. The majority are grown on relatively small farms with ten or fewer employees but there are some larger companies. That situation is changing as farms increase in size.

A different species, Pacific oysters are grown on the West Coast. Most of the large-scale companies are located in Washington where many of them actually own intertidal land rather than lease public land. A number of them are vertically integrated companies that operate from the hatchery to the consumer, including wholesale and retail sales, as well as restaurants. The comparison between the coasts is significant, although more East Coast companies are also integrating operations from the hatchery to the restaurant and consumer.

Questions remain about shellfish aquaculture along the coast. Will there be suitable space for continued expansion? Will the public embrace shellfish aquaculture, and will aquaculture be compatible with recreational and commercial uses of the water? Will communities and states invest in a public effort to maintain long-term shellfish culture? How will climate change affect the industry, and can the industry adapt to the changing environment? Will the industry move to sites offshore instead of dealing with the multiple uses of inshore areas?

The rapid rise in shellfish aquaculture has increased its visibility as a viable industry. It has also raised awareness of another potential benefit. Some feel that shellfish, as filter feeders, may help to address the wastewater problems that have impacted the Cape and our coastal areas nationally. And so, if there is compatibility with other competing uses of our coasts, will shellfish aquaculture be promoted as a positive way to help mitigate the negative effects of nutrient loading? Will it be accepted as a green industry that produces food, jobs, and societal ecological benefits?

# Chapter 21
## Shellfish to the Rescue?

One summer day in the early 1980s the water was an odd rust color in Pleasant Bay's Meetinghouse River. A closer look at a sample under a microscope revealed that the plankton was almost exclusively one species. It was an unusual monospecific bloom, as it's called when one species dominates the plankton community. This bloom was concentrated highly enough to color the water. That was not good news.

Several years later, roaring seawater cascaded into the tanks at Orleans's shellfish propagation facility nicknamed simply "the lab." Pumped from the Town Cove in a flow-through seawater system, the water carried essential phytoplankton to feed a million baby seed quahaugs in the upwellers. Yet the tiny animals had stopped feeding. Instead of thousands of opaque little siphons sticking out from slightly parted shells, constantly taking in the life-giving water, the animals were shut tight in every silo in every tank.

There was plankton in the water, but almost all of it was a single species of algae that had sharp spikes, a trait that quahaugs apparently did not like. When that species, or one like it, is in the water, the quahaugs shut down and don't feed. It was late July, water temperature was high, and oxygen was low when this spiky species appeared.

Seed quahaugs do not survive long without good food, especially in a reduced-oxygen environment. Suspended feeding activity by the tiny seed for an extended time is never a good thing.

The algae species was nontoxic, but it was a close relative of *Alexandrium tamarensis*, the toxic species that causes paralytic shellfish poisoning, or "red tide," in the spring. The red tide organism has two phases of its life: a "cyst," or seed, and the floating plankton. When conditions are right, the cyst germinates and transforms into the plankton, which quickly multiplies into a bloom. Shellfish feed on the toxic plankton, concentrating the toxin in their tissue. When we eat the shellfish, that toxin affects our central nervous system. If severe enough, it can paralyze our respiratory system, hence the name "paralytic shellfish poisoning." States monitor the water carefully and ban shell fishing completely when the red tide is present in dangerous concentrations. The red tide does nothing to the shellfish, however, and when the bloom disappears and the plankton transform to the cysts that drop back down into the sediment, the shellfish purge the toxin from their tissues over time and become safe to eat again.

Alexandrium is one of many types of toxic plankton that create harmful algae blooms. Those monospecific blooms of algae are increasing worldwide, partly because of increased nutrients in the seas and estuaries. Both species, the nontoxic form and the toxic "red tide" species, belong to a group of species called "dinoflagellates." While the species affecting the quahaugs was not toxic, it did indicate an unbalanced plankton community. That reddish water meant that an elevated level of nutrients were fertilizing the plankton. This was also not good news. It indicated that our activities on land were affecting the water.

In the late 1990s, the Cape Cod Times published a photo of seaweed along the shore of Waquoit Bay. It was startling because the seaweed was thigh deep and extended at least twenty feet from the edge of the shore. It was one more piece of proof that nutrients from the land were seriously affecting the estuaries.

Signs of malaise in the Cape's estuaries were everywhere – clams and mussels covered by green filamentous seaweed; banner scallop harvests in 1975, 1983, and 1985 in Orleans and then nothing significant afterwards, coinciding with declines in eelgrass and a similar lack of scallops all over the Cape and southeast Massachusetts; monospecific blooms of phytoplankton; deep seaweed mats lining the shore; shellfish habitats deteriorating to heavy organic, smelly mud. People were beginning to notice that all was not well.

Ever since indoor bathrooms have replaced outdoor privies, Cape residents and millions of visitors have flushed their human waste and household products down the drain. That waste is loaded with nutrients as well as harmful bacteria from our bodily systems, toxic substances, and pharmaceuticals, a new threat. Now, Cape Cod faces its greatest environmental challenge ever – what to do with our wastewater and how to manage it.

Nitrogen, a nutrient generally associated with its beneficial qualities as fertilizers for gardens, gets mixed into the groundwater every day from the hundreds of thousands of septic systems on the Cape. The nutrient-loaded groundwater flows steadily into the estuaries. For estuaries, too much of a good thing is a bad thing. The result of estuarine over-fertilization is called nutrient enrichment or eutrophication. About 80 percent of the controllable nitrogen causing the issues in our estuaries comes from our septic systems. Farming practices are often the primary source of the excess nutrients in other parts of the country. People cause the problem wherever it may exist, not nature. It will not go away on its own. It will only get worse.

Traditional septic systems eliminate the bacteria in our waste, but not the nutrients. The two most significant nutrients are nitrogen, which affects saltwater, and phosphorus, which affects freshwater lakes and ponds. The effluent from septic systems mixes with the rainwater and snowmelt that becomes groundwater. All of the water beneath our feet moves outward from the center of the Cape and eventually reaches the nearest bodies of water.

The Cape's island geography dictates the direction of groundwater flow. If a dividing line were drawn on a map down the center of the Cape from Sandwich to Brewster, it would illustrate the direction of the flow because the water moves perpendicular to the line – north toward Cape Cod Bay or south toward Nantucket Sound. Similarly, a line drawn down the center of the towns fronting Cape Cod Bay from Eastham to Provincetown has water flowing toward Cape Cod Bay or the Atlantic Ocean. Since some of the land in Bourne and Falmouth fronts on Buzzards Bay, some of the groundwater from those towns flows into that bay. The line between one direction of flow and the opposite direction delineates a watershed. All of the land within the dividing lines contributes groundwater to a specific bay. That is a general overview, but the full story is much more complicated.

Smaller estuaries – harbors, bays, coves, and marshes – intercept the groundwater before it reaches the larger bays. The negative impact from nutrient loading is most acute in those estuaries. Each separate estuary has groundwater flowing into it from the surrounding land; dividing lines from one to another delineate smaller watersheds. Determining and mapping the watersheds, many of which are complex, is the first step toward planning for wastewater management.

Orleans and Chatham have some of the most complicated watershed delineations. In Orleans, groundwater flows into either Cape Cod Bay or the Atlantic Ocean. But before the groundwater gets to the larger bays and ocean, it flows into marsh creeks on Cape Cod Bay, or into Nauset or Town Cove, or Pleasant Bay. The latter three have numerous offshoots – salt rivers, ponds, and coves – called sub-watersheds. Each sub-watershed has its own delineation lines which indicate where the groundwater flow changes from one sub-watershed to another. Thus, when trying to figure out exactly what is happening and how to mitigate the effects of wastewater on a particular pond, or river, or larger area, it is important to understand the relationship of the sub-watersheds. That watershed delineation determines which individual properties contribute wastewater to which watershed and where the nutrients from each bathroom are heading.

Chatham's watersheds are similarly complex. Some empty into Pleasant Bay and eventually the ocean, while others empty into Nantucket Sound. Multiply Orleans and Chatham watershed delineations by all of the sub-watersheds on the Cape, and you can see it is an extremely complicated situation.

Planners have broken the problem down into more manageable individual watersheds. They mapped all of the septic systems that could flow into a particular body of water and determined where the groundwater flow changed direction from one watershed to another. They added residence times – the time it takes water to flow from the estuary to the sea – to more precisely determine the wastewater issue at each watershed. The end result was a series of maps for each watershed that indicate the areas most vulnerable to the effects of excess nutrients. That information could be combined to understand the entire picture and

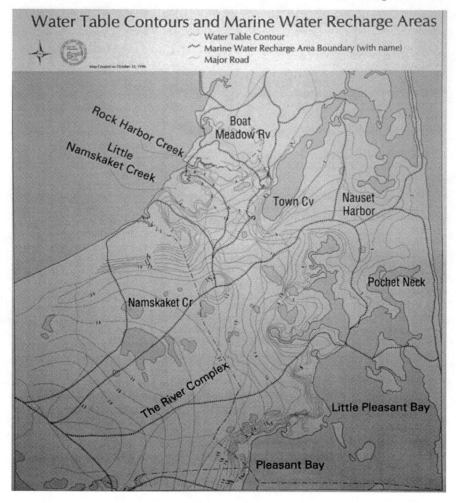

**Primary watershed delineations in Orleans. There are also numerous sub-watersheds that complicate accurate mapping and the planning process. Groundwater flows perpendicular to the delineation lines. (Cape Cod Commission)**

determine the best solutions.

It is difficult for many residents to comprehend that the effluent from a septic system far inland affects the health of estuaries miles away. It's also difficult to comprehend the speed at which the effluent travels. Generally, groundwater travels about a foot a day in most Cape soils. Therefore, what is flushed from a toilet to a septic system located a hundred feet from an estuary, the minimum distance allowed by regulation, will take one hundred days to reach the estuary. For three hundred feet, it takes almost a year. For a mile, the time lag is over 5,200 days, or about fourteen years. Effluent from houses built near the water during the building boom in the 1970s started affecting the estuaries quickly. Houses built in the woods or inland commercial or residential developments constructed far from the water did not affect estuaries as quickly, and their contribution to the problem was not discernible for a long time. As building accelerated during the final quarter of the twentieth century and into the twenty-first century, the problem magnified as the concentration of nutrients increased – new effluent being added to what was already in the ground. The increases continue unabated except in towns like Barnstable and Falmouth that have sewer systems in some parts of the communities. Chatham has adopted a comprehensive wastewater treatment plan that is expected to encompass two-thirds of the town by 2030.

Estuarine systems adjust incrementally to increased nutrients. Eventually, they can no longer assimilate the excessive nutrients and contaminants that bombard them for years. Then they change visibly. Heavily concentrated plankton blooms change the color of the water or turn it murky, which blocks sunlight to eelgrass. Mats of seaweeds build up along the shore. The plankton and seaweed use oxygen while they are alive. After they die, the decay process eats up oxygen as well. Overall, the oxygen level of the estuary drops below essential levels and water quality declines. Eelgrass disappears. Fish kills become more common. Shellfish habitat is lost as the friendly sediment of sand, silt, and shells become smelly, heavy organic mud that is inhospitable to shellfish. Scallops and other shellfish populations disappear or decrease. Meanwhile, as home and business construction continues unchecked in numerous locations, the additional septic systems exacerbate the problem.

Crucially, addressing the wastewater problem does not immediately fix the problem. Effluent will remain in the groundwater for decades simply because it is already there, flowing toward the estuaries with no means to capture it along the way. What can be done about this problem?

～≋≋～

Recognition of the dire circumstances statewide led regulators in the Massachusetts Department of Environmental Protection to initiate the Massachusetts Estuary Project. They contracted with the University of Massachusetts Dartmouth's School of Marine Science and Technology (SMAST) in December 2001 to assess each estuary by modeling the flow characteristics and residence times. Their report provided technical data that identified the sources and amounts of nitrogen entering the sub-embayments. The report for each major watershed – Pleasant Bay, Nauset, Chatham's Stage Harbor, Barnstable's Three Bays, Waquoit Bay in Mashpee and Falmouth, Popponesset Bay in Barnstable and Mashpee, and Bass River in Yarmouth and Dennis – vividly depicted the nutrient problem. Each report included a map of where the nutrient concentrations were the greatest and least. In most cases, the highest concentrations were in those areas with the longest residence time, the amount of time water took to get from any point in the estuary to the ocean, color-coded on the maps.

Changes sometimes occur incrementally, and each incremental change becomes the norm so subtly that we tend to forget what the new norm replaced. In 1977, when we constructed sand-filled floating rafts for a seed quahaug nursery system, we anchored two rafts in Lonnie's Pond in the upper reaches of Pleasant Bay and two in the Mill Pond in Nauset. The quahaugs consistently grew faster in Lonnie's Pond. After four years, seaweeds were floating in Lonnie's Pond but not in the Mill Pond. Chicken wire covered the tops of the rafts to prevent predation by birds, but we left the ends open so water could flow freely. Seaweeds floated into the rafts and began to smother the quahaugs, so the rafts had to be cleaned out regularly. Different types of green seaweed attached to the rafts' styrofoam floats, adding weight and sinking the rafts deeper in the water.

At the same time, huge amounts of bushy, red filamentous seaweeds lined the shore of the pond where we had been transplanting some quahaugs from the rafts at the end of the growing season. The transplant sites had changed to a mucky substrate. By 1983, we could not keep up with the maintenance required to keep the rafts at Lonnie's Pond, and we could no longer use that site as a transplant area either.

The 2009 maps of Pleasant Bay produced by SMAST showed that Lonnie's Pond was seriously impacted by nutrients, and they provided another map that predicted heavier nutrient loads in the future. With a combination of future building and Chatham inlet migration farther and farther away from the pond, the outlook was grim. Mill Pond was less impacted but was showing signs of decline and was also subject to inlet migration. My observations from two decades earlier were confirmed. Nutrients from the land flowing into the ponds was a serious problem.

The Department of Environmental Protection used the SMAST report to set the maximum daily amounts of nitrogen acceptable for each Cape estuary, with the goal of protecting its character and use. When the nitrogen concentration exceeds the total maximum daily load, the term used by the state to determine how much nitrogen a water body can assimilate to function properly, the town is responsible for reducing the amount of nitrogen flowing into the estuary to acceptable levels. The state contended that when nitrogen reduction was implemented and well-managed, the estuary's health would be restored and it would be protected from further degradation.

As the planning process progresses and Cape Cod inches closer to implementing corrective measures, the question of cost – not necessarily the end product of cleaner water – has become the focus. Complicating the problem is that many Cape estuaries are shared by two or more towns. How they should share the responsibility for nutrient mitigation and cost is difficult to resolve.

Anxious about saving money, groups of citizens questioned the validity of the SMAST researchers' findings. They debated the basic science behind the assumptions, the methodology employed, and the modeling used to set the load levels for each sub-embayment of each estuary. Questioners challenged the amount of nitrogen flowing from groundwater and the relationship between human-derived nitrogen, mainly from septic systems; and from natural sources, precipitation and the amount flowing into the estuaries with the tides. They cast doubt on the official reports. That doubt led to a peer review of both the methodology and science. The review found both to be sound.

The primary detracting argument was that the amount of contaminated groundwater and nitrogen from the residents is overstated when there is already so much nitrogen coming into the system from natural sources – precipitation and tides. We know that there is generally more nitrogen added to the estuaries by atmospheric deposition and tidal influences than by people. With just these natural sources, though, the estuaries function well. Estuaries adapt to changing conditions until a tipping point is reached. In this case, the nitrogen from everyone's septic systems being added to the groundwater flowing to the estuaries is the unnatural condition. We have tipped the scales, putting the estuarine systems over the edge and out of balance. Our contribution is the only component that can be controlled. We have caused the problem that only we can resolve and restore the estuaries.

Concern surfaced that a county mega-agency modeled after the Massachusetts Water Resources Authority (MWRA) might be created. That public authority was created in 1984 to clean up the polluted Boston Harbor. Responsible for a gargantuan system, MWRA provides water and sewer services to sixty-one metropolitan Boston communities, covering 2.5 million people and more than 5,500 large industrial users. While county officials have refuted the claim that they intend to establish such an agency, most towns have balked at implementing solutions that have been presented thus far.

Chatham is the exception. The town has installed sewers throughout much of the community, accomplished because it is one of the few places on the Cape to already have a sewage treatment facility. Chatham also acknowledges the importance of its fishing industry, both finfish and shellfish, and residents were quick to recognize the wastewater issue. Their foresight was rewarded with substantial funding from low-interest loans to fix it, but they also voted to spend their own funds at a substantial cost to themselves. Other towns less equipped to deal with the monumental hurdles of constructing sewers face tremendous expenses, the prospect of major capital construction, substantial disruption during construction, and significant continual operation and maintenance costs.

The crux of the issue is how to reduce the nitrogen and at what cost. Funding mechanisms remain elusive. Barnstable County residents, faced with paying the entire multibillion-dollar bill with no firm respite in sight, are understandably concerned. The Massachusetts Department of Environmental Protection lacks the finances to assist in correcting the problem, and federal financial assistance appears limited or nonexistent. State legislators and congressmen alike are cognizant of the situation, and they recognize Cape Cod as a national treasure. But they have not figured out a way to take the full financial burden off individual homeowners. At this point, each community is left to fund millions of dollars for wastewater projects on their own – a scary prospect that is difficult to achieve.

For long-term Orleans residents, the current debate over sewer installation is a case of déjà vu. In the 1980s, Massachusetts forced towns to stop dumping waste from backyard cesspools and septic systems into unlined open lagoons. The choice then was to construct a sewage treatment facility plus a sewer system for the downtown district or construct a septage-only treatment plant for "solids" from the backyard systems. The combination option of treatment plant and sewers would effectively stop most of the nitrogen from entering the groundwater, which was known to flow into Town Cove and Cape Cod Bay. Either option, sewers or septage only, would have been substantially financed by state and federal funds. After much acrimonious debate, numerous committee reports, lost friendships among neighbors, and many votes, the town approved the septage-only option instead of sewers.

Now, almost forty years later, Orleans faces the same decision about constructing a sewage system for the downtown area as part of a comprehensive, six-phase plan. Only this time, there is no government money to underwrite the costs. Orleans began planning for wastewater management in 2000, with a steering committee consisting of representatives from the Board of Selectmen, Board of Health, Conservation Commission, Planning Board, water commissioners, the Finance Committee, and residents at-large. The townspeople voted for the proposed Comprehensive Wastewater Management Plan. It included a centralized treatment plant and sewers for the center of town, alternative wastewater treatment methods for other areas, and comprehensive reviews before each phase was implemented.

But between voter approval in 2008 and when the plan was approved by both the county and state in 2011, groups of citizens questioned the plan's validity, the Massachusetts Estuary Plan on which it was based, the options selected for implementation, and the cost of implementation. Opponents presented and defended alternative, less expensive options, maintaining they were equally effective. An independent review found that while there would be a small savings with the substitute plan, there were also significant yearly maintenance costs, and effectiveness of the suggested options was also challenged.

At its annual town meeting in May 2013, Orleans faced the first funding article, $3.5 million for the first phase of planning and design that required a two-thirds majority to pass. The article was defeated by six votes in one of the largest town meetings on record of over nine hundred people. Proponents next delivered a petition for a special town meeting, signed by over six hundred residents. This time, over fourteen hundred people poured into a sweltering school gymnasium, the town's largest turnout ever. The article again garnered a majority of votes but failed to reach the required two-thirds.

For many, the initial questions about the Massachusetts Estuary Project and then the questions surrounding the town's wastewater management plan after over a decade of public meetings and planning, were seen as delaying tactics aimed at avoiding spending any funds, or prioritizing cost over effectiveness of the solution. To others, the questions were legitimate concerns to ensure that when funds were approved, it was at the least expensive option and everything had been completely vetted. Both the effort in the eighties and the most recent episode had the same side effect:

dividing the community. The earlier rift lasted for a long time and was at the heart of many substantial changes in how the town operated. The later version will also affect people for many years. Finally, however, the first financial installment was approved in May 2017, four years after the six-vote loss. Millions had been spent to address concerns in order to begin the long journey toward addressing the problem about nutrients.

As people across the Cape grapple with similar scenarios, they search for ways to ease the financial burden posed by building massive wastewater treatment facilities. The example of Barnstable County in general and Orleans specifically has been repeated in coastal communities elsewhere. The problem is well-known. The solution is much more elusive. Some wonder if shellfish cultivation could be part of the solution.

Shellfish, as important members of the estuarine ecosystem, have traditionally been extremely important on the Cape and in most coastal communities. They provide food, jobs, and an economic boost to the communities. Shellfish also filter water that reduces turbidity, increases light penetration that stimulates eelgrass growth, and improves water quality. They stimulate bacterial action that chemically reformulates nitrogen compounds. Oyster reefs increase biodiversity and add height and structure that can reduce the effects of waves. Combining all of these factors, shellfish provide societally beneficial "ecosystem services."

Given the depletion of shellfish in the estuaries, there is a strong desire to replenish them, but funding shellfish propagation on the Cape is not usually considered an essential municipal service in cash-strapped towns. Now, some people are viewing shellfish differently – as a lower-cost option to clean up the estuaries. The suitability and efficacy of individual species of shellfish for cleaning up the estuaries has come under scrutiny.

The nature of mussels – their tendency toward gregarious setting, their natural productivity that makes them the most prolific of the local shellfish, their feeding behavior that enables them to digest dinoflagellates, that group of phytoplankton species disdained by other species such as quahaugs – suggest that they would be good for nitrogen reduction.

Unfortunately, they prefer high salinity water and are not found in the upper reaches of estuaries where most of the problems associated with nutrient enrichment occur.

Clams, which also exhibit gregarious setting, are found in a wider range of salinities, but they are not as efficient at pumping water and, therefore, fall somewhere in the middle range of candidates for nitrogen removal. Quahaugs are not as prolific as softshell clams or mussels, do not tend toward gregarious setting, grow slowly, and live in wide salinity ranges; and their pumping rate is considerably slower than oysters or mussels. However, they live longer, a trait that may ensure a more stable population than short-lived species. It is still a numbers game, though, because there must be enough parent stock to sustain the population.

Scallops naturally set so sporadically that culturing them is the only real way to maintain a stable population. Aside from the Martha's Vineyard Shellfish Group, few other places or individuals culture them, as they are such a delicate creature.

That leaves oysters. Oysters have been functionally extinct throughout the Cape for generations except in Wellfleet. However, historical shell middens, the dumping grounds for discarded shells, reveal a vastly different story of high abundance during the centuries that oysters provided sustenance for the Wampanoags all over the Cape. Towns have investigated methods to bring them back as part of their propagation responsibilities, and profitable private oyster aquaculture has mushroomed statewide during the past two decades. Oysters can pump a lot of water, up to fifty gallons a day in some locales, which makes them a good candidate for this purpose.

Young shellfish such as seed oysters must be protected from predators and competitors. That means using aquaculture gear – plastic net bags used to raise young shellfish in a protected water environment. Aquaculturists, including towns using aquaculture methods for propagation, employ a variety of gear. For oysters, they use mesh fashioned into bags fastened to rebar racks; vinyl-coated cages; floating bags suspended on the surface; mesh bags filled with spat-on-shell, tiny oysters that have attached to larger shells; or various combinations that protect the shellfish from predation.

Oysters live on the surface of sediment, and if they sink into soft mud, they suffocate and die. "Hardening" the muck requires tons of shell or pieces of concrete or other similar material.

Quahaugs and clams require the right type of sediment to grow properly. Growers cover wide swaths of tidal flats with plastic mesh netting. Mussels require something to attach to and are often cultured on ropes suspended in deep water.

A combination of species that would more closely resemble natural conditions would be preferable to monocultures because all species are filter feeders. Studies are currently underway to determine exactly how efficient various shellfish species, including some non-commercial species, perform in nitrogen remediation. But implementing a culturing program with shellfish of different environmental requirements adds tremendous complexity to the process.

Regardless of whether shellfish are cultured by a municipality in a public propagation project or commercially, some sort of aquaculture gear is essential. And the gear takes up space that may conflict with other uses of the water. If the shellfish are grown in totally subtidal waters, there may be little problem to boaters, depending on the water's depth. Careful boaters will not get tangled in netting; shellfish leases are well marked and easily spotted. However, intertidal gear is visible when the tide ebbs. Floating bags that are connected by ropes to buoys that are anchored to the bottom are more difficult to avoid and are visible at all times.

Both intertidal and floating gear have become a "cause célèbre." Shorefront property owners argue that the visible aquaculture activity mars their scenic view, lowers property value, and is generally unsightly. The classic "Not in My Back Yard" attitude often prevails. However, the waters are public domain and even private aquaculture takes place in leased public waters. Scenic views are important, of course, but private property owners dictating policy because of the impact of aquaculture on their exclusive view of public waters is a very different matter.

All over the country, researchers are evaluating the efficiency of shellfish to mitigate nitrogen. Orleans began a pilot project in Lonnie's Pond in 2015 to answer the question specifically for the town. The project has been expanded each year and results so far mirror results elsewhere – oysters are efficient at removing significant amounts of nitrogen. There are other considerations though.

**Pilot oyster project, Lonnie's Pond, Orleans. Oysters in floating bags are being tracked to determine the amount of nitrogen removal from the pond by the oysters. (Photo by author)**

Storm water provides a pathway for bacterial and other contaminants flowing in groundwater and surface water to foul the estuaries. Other fairly new concerns are pharmaceutical compounds and frequently-prescribed hormones which enter the groundwater after unwanted medicines are flushed down the drain as well as residual amounts in human waste. If or when they reach the marine environment, their effects on the marine system, and efficient treatments to remove them from the effluent waste stream before that happens, are important emerging courses of study.

The porosity of the soils leads to unintended consequences when a toilet or kitchen sink is used to dispose of expired pharmaceuticals and household products. Such substances can be highly toxic in the marine world, and one can only imagine all of the pathways that these toxic elements might follow. Moreover, the effects of these potential toxins could come back to haunt us through drinking water. It is doubtful that shellfish can solve that problem.

The irony is that shellfish need clean water to be harvested, and aquaculture fits into this picture. Shellfish aquaculture is a green industry that also partially counteracts the effects of nitrogen introduced by tourists, seasonal residents, and year-round people whose presence impacts the waters so drastically and for such a long time.

Shellfish culturing may sound like a less expensive solution than the major infrastructure necessitated by a sewage treatment plant and the sewer system required to carry effluent to the plant. It is impractical, though, for shellfish to be the primary method for eliminating enough nitrogen to meet the total maximum daily load levels. It would require raising millions of shellfish to start the process and then doing it every year for decades.

Shellfish at the end of the effluent trail do nothing to get to the root of the problem – the one that starts in the bathroom and the backyard septic system. Adding more shellfish will maintain or increase biodiversity and is societally beneficial, providing desirable ecosystem services. And they can reduce nitrogen levels.

But the addition of shellfish alone will never eliminate nitrogen pollution. Filling the estuaries with shellfish through town propagation programs, private aquaculture, or a combination is not the entire answer to wastewater management. Shellfish culture can be a valuable component of an overall program but one cannot simply obtain and place seed in protective gear and expect and rely on successful growth or survival for years into the future. Culturing shellfish means understanding and dealing with natural systems, realizing that things can go very wrong at any step in the process and seriously impact or destroy a year's worth of effort or worse.

There are obstacles: cost; seed availability and reliance on hatcheries to provide the seed for public and private needs; reliance on stable, long-term funding to support the culture including labor and nursery culture infrastructure; reliance on private businesses that lease public water to be long-term partners; natural environmental changes such as ocean acidification, and the unknown consequences of climate change, to name a few. Add to that the amount of public space normally used for other competing activities that would be required to raise the amount of shellfish annually. With all these considerations, the decision of whether to pursue this route becomes even more difficult.

There is no question that estuaries filled with enough shellfish to resemble earlier periods of abundance would be much healthier systems. Shellfish are a great benefit to society. However, we need to address the land-use issues, especially those involving our bathrooms at the beginning of the waste stream while we let shellfish address the issues at the end of that stream.

We are asking a lot of shellfish. Once again, we want it all. We want them for food, for jobs, for recreation, and for ecosystem services. And now we want them to clean up the water that we degraded so much that it caused their habitat to decline in the first place. But shellfish are not a panacea for wastewater management.

The questions remain: Can we have shellfish populations that continue the centuries-old fishing traditions? In the process, can shellfish help save the estuaries? Can we afford to have balanced estuaries functioning as they are intended? Can we afford not to have them?

Chapter 22
# An Oceanic Conveyor

The web of life teaches us about the interdependence of biological systems. We can grasp the concept of predator and prey dynamics and the food chain hierarchy of animals from producers to consumers to top consumers. We seldom think that the non-biological part of our world also exhibits interdependence or how much the biological component depends on the physical one. In that realm, the importance of the Gulf Stream that transports more water than all the world's rivers combined, according to the National Oceanic and Atmospheric Administration, cannot be overstated.

Climatologists contend that what is known as the North Atlantic Ocean conveyor is the primary engine that drives a continuous global loop of circulating warm and cold ocean currents that govern Earth's climate. Furthermore, researchers have determined that it is slowing down. The consequences will be felt around the world. One person's journey across the Atlantic illustrates the complexity of what is happening with the Gulf Stream.

Sarah Outen was 29 in May 2015 when she left Chatham, Massachusetts, to row across the Atlantic in a twenty-one-foot, double-ended rowboat. She anticipated being in that small craft for about ninety days before she made landfall in England. Her boat, *Happy Socks*, had two covered compartments, one in the stern for sheltered living accommodations and one in the bow for supplies and gear. In between

Sarah Outen all smiles aboard her boat *Happy Socks* as she gets ready to begin her journey across the Atlantic. (Courtesy of Sarah Outen)

Sarah leaves Chatham on May 14, 2015 to complete the final leg of her epic journey to circumnavigate the globe on human power: kayak, bicycle, and rowboat. (Courtesy of Sarah Outen)

was the open cockpit with its sliding seat to accommodate her hundreds of thousands of strokes. Two strong, black, two-inch-thick crossbars connected to the boat's port and starboard sides between the covered compartments were her only obvious safety measures to keep her inside the cockpit amidst steep Atlantic swells and mountainous waves.

Sarah had arrived in Chatham a month or so earlier intending to complete the grueling odyssey of becoming the first person to circumnavigate the earth propelled only by human power – by kayak, bicycle, and rowboat. After kayaking down the Thames and across the English Channel, she biked across Europe and Asia and rowed to Japan. A fierce typhoon forced her to abort her attempt to row across the Pacific. She was rescued by Japanese authorities. Scared and humbled, she took a couple of years to regroup, then rowed across the Pacific, kayaked the Aleutians, biked across North America during a punishing and unrelenting winter, and arrived in Chatham to prepare her boat and body for the last leg of her epic voyage – rowing three thousand miles across the Atlantic and back to the Thames for a celebration.

One day during her month of preparation, I had lunch with Sarah and

Marj Burgard, a world-renowned senior rowing champion. That day's *Cape Cod Times* was on the counter near the restaurant's entrance. The front page had a map of the North Atlantic depicting the Gulf Stream's location based on sea surface temperatures. The picture took me by surprise. The Gulf Stream was not located where I thought it would be or looked like I imagined it would. It was much farther south and east. It also appeared to be less of a defined "stream," and more of an amorphous mass covering a large portion of the North Atlantic, with several areas that looked like curled offshoots of current. I looked at the map with a sense of foreboding about Sarah's journey. She would have to row much farther east before she reached the majestic current that would help carry her across the sea. And the current did not seem to be behaving as expected. I hoped that by the time she needed that flowing ocean river of warm water, it would be where it could propel her toward England.

For centuries, the Gulf Stream has transported ships across the Atlantic. Benjamin Franklin was the first to name the current after noticing particular seaweeds far out to sea and noting that mail was delivered faster from America to England versus the opposite direction. In addition to its speed, the Gulf Stream raised the average water temperature ten to twenty degrees Fahrenheit. That temperature differential warms the eastern side of the Atlantic, making northern Europe, especially the British Isles, the North Sea, and the Baltic much warmer than areas of equivalent latitude on the western side. London is farther north than Quebec. Exploiting the fisheries and marine mammals in the North Atlantic was possible because of the mix of warm and cold water combined with deep and shallow bathymetry, creating conditions for tremendous productivity and diversity. Yet information from NOAA's National Climate Data Center indicate troubling shifts from the norm.

The oceanic conveyor of the Gulf Stream works because it is like an "up" escalator. When an individual step of an escalator gets to the top, it invisibly goes back down to the bottom via the underlying mechanism to begin the journey again in a continuous ellipse. The Gulf Stream behaves in a similar fashion. Warm water, the bottom step, travels north from the Gulf of Mexico, through the Straits of Florida, up the East Coast, and veers northeast somewhere off Cape Hatteras in North Carolina. The current splits when it reaches Europe – south, along the coast of Spain and Portugal, and north, where it splits again toward the British Isles, Norway, and Iceland. It reaches the top step of the escalator in those colder, higher latitudes. It loses some heat as it meets cold Arctic currents flowing along Labrador and Greenland.

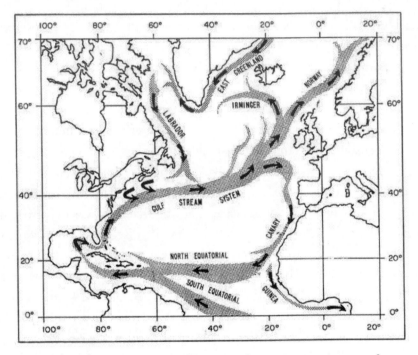

**North Atlantic Currents where southern-most warm surface currents transition to cold deep water currents.**
**(Photo taken from Pinterest, Public Domain)**

As the Gulf Stream's warm water cools, it becomes heavier and denser and sinks, forming the back side of the escalator mechanism as it flows down to the bottom again. The cold water flows south in the deep water off the continental shelf of North and South America toward Antarctica. Scientists call this the Atlantic Meridional Overturning Current, indicating the latitude meridians that the current crosses on its journey. The AMOC splits near Antarctica, part of it heading toward the East Atlantic and the other part toward the Pacific. Part of it gains heat as it travels toward the tropics and flows toward the surface to begin a different loop of warm to cold depending on latitude. Eventually, the AMOC gets back to the Gulf of Mexico.

That movement of warm tropical surface water that flows toward colder water, cooling, sinking, and traveling along the bottom in the deep water, acts as a vast ocean conveyor. Unlike an escalator's elliptical cycle that takes less than a minute, the oceanic conveyor's journey around the earth takes an estimated one thousand years.

Sarah left later than she had hoped. Uncooperative easterlies caused delay after delay, but finally, on May 14, her supporters and new American friends bid her good-bye and safe travels as the Coast Guard accompanied her through Chatham Inlet and she got on her way.

After a couple of days of westerly winds, Sarah was plagued again by easterlies, continually pushing her backwards toward land. She had to escape their grip and get near the Gulf Stream, usually several hundred miles east of the Cape. Finally, she got the push she needed from favorable winds and headed northeast toward Newfoundland. When she got there though, she spent several days rowing in an enormous circle. Ever upbeat, she accepted the challenge and interpreted it as a life lesson.

The Gulf Stream, often described as a river of warm water, is not like a classical river. There are no earthen banks to keep it penned in along a certain path. It has no solid sides or defined streambed. It is fifty to sixty miles wide, two thousand to three thousand feet deep, and is the fastest current on earth, flowing at about 4.5 miles per hour while transporting twenty cubic miles of water a day. It is known to shift position thirty to sixty miles north or south, and often the location varies with the seasons.

A land-based river flows swiftly until a large obstacle like a rock or a tree trunk, perhaps lodged between rocks, diverts the water around the obstacle, eventually creating a meander. It may take decades or even centuries for the river to eat away at the bank to create a meander, which eventually may close while the river finds a new path.

The Gulf Stream also meanders, but these diversions can be fleeting aspects of the flow. Some of the meanders form rings – warm water core rings along the northern boundary or cold water core rings on its southern edge. These rings are one hundred to two hundred miles in diameter and are critical for transferring heat. The northern warm rings generally move in a southwesterly direction before they are re-absorbed by warmer tropical waters.

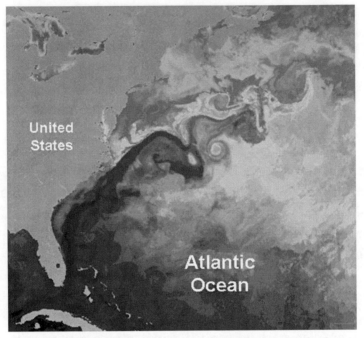

**Sea Surface temperature indicating Gulf Stream in darker tone.
Note meanders and eddies breaking off main current.
(National Oceanic and Atmospheric Administration)**

The existence of warm and cold core rings had potential implications for Sarah. It represented a dot like the ones in a connect-the-dots drawing, just as the "offshoots" and amorphous warm water did in the newspaper map.

At about the time I met Sarah for lunch, NOAA's Climate Data Center released a report stating that the period from December 2014 through February 2015 was the warmest on record for the entire Earth, even though the winter in the Northeast was among the coldest. Another anomaly was that the winter produced record cold water temperatures in the same part of the North Atlantic where Sarah was headed a few months later that spring. The report included a color-coded map depicting average temperatures: shades of red for warm and shades of blue for cold. The Atlantic east of Cape Cod and south of Nova Scotia and Newfoundland was colored in shades of red. But south and east of Greenland was a large block of blue, very cold water. It was odd because most of the earth was shaded in red.

As unusual as that map was, the placement and type of cold water have broad implications for the Gulf Stream.

Scientists have been tracking the strength and location of the Gulf Stream. Researchers have found that climate change from greenhouse gases is altering the previously normal and stable system and that the AMOC has been weakening since the mid-seventies. At this point in our understanding, the possibility of an abrupt collapse appears improbable for this century, but estimates indicate a 25 to 30-percent decrease in the Gulf Stream's strength during the twenty-first century.

Warning signs are evident. The global warming effect we are experiencing now, most evident at the poles with their decreased ice cover, is one of those signs. NOAA released a video in 2016 that shows time-lapse images of the Arctic ice melt from 1990 to 2014. That video vividly illustrates a tremendous loss in "old" ice that has been constantly present, over the fourteen-year timeframe. Researchers have reported huge amounts of freshwater entering the sea because of the melting. The video shows the current flowing from the Arctic polar ice cap to the east side of Greenland, although some also flows to the west.

Scientists have also noted that Greenland's ice sheet is melting more rapidly than suspected just a few years ago. The melted water from the polar cap and Greenland is less dense than saltwater and floats on top of the higher density cold seawater. When the influx of cold freshwater meets the remnants of the Gulf Stream's warm surface seawater, the two water masses both float on top. Some mixing most likely takes place, forcing the heavier warm saltwater to sink. But if that is not the case, the floating cold freshwater might "block" the warm Gulf Stream seawater from following its normal route to where it will cool and sink. If the Gulf Stream can't move north, the warm water "backs up" in warmer latitudes and the current slows down.

**Unusual "Cold Blob" south of Greenland. The anomaly has persisted for several years, including into 2018, while rest of globe was warmer than average.**
**(Photo courtesy of AccuWeather)**

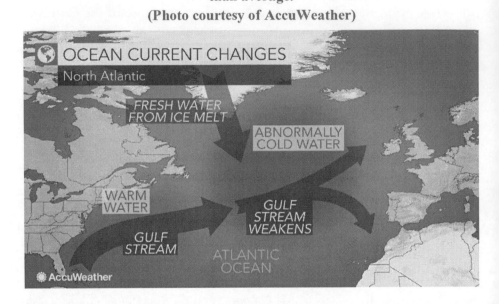

It appeared that Sarah's navigational troubles were signs of a much more pronounced problem that affected a great deal more than her personal journey. The signs began far earlier than her departure in May. They had become increasingly evident in the North Atlantic for several decades. Climatologists around the world have been monitoring the conditions, but what is happening in the North Atlantic – the slowing Gulf Stream – has just begun to register with the general public.

As I looked at the maps, videos, and information from NOAA, I wondered whether that large block of warm water filling the area from Cape Cod to the mid-Atlantic was what I had noticed in the *Cape Cod Times's* graphic image. Was that wide swath of water really the Gulf Stream instead of the fast-running "river" I had been led to expect? If the block of cold fresh water was meeting the warm Gulf Stream seawater and both were near the surface, was the ocean turbulent enough to create eddies?

Was the effect similar to what happens when air masses of different pressures collide and produce high winds? Were the "offshoots" I observed actually warm water core rings that were forming? Had Sarah and her little rowboat been distressingly caught in one or more of those giant circles that were moving southwest, back toward Cape Cod? It seemed as if her boat was more like a carousel horse going from one giant oceanic merry-go-round to another instead of being swept forward by the previously predictable warm-water river that was the Gulf Stream. Sarah, apparently, never found that elusive, majestic river.

While the warm Gulf Stream moderates the temperatures in Europe, large climatic forces add complexity. One of the forces, called the North Atlantic Oscillation, exists because of seesawing fluctuations of an area of high pressure at the Azores and a low pressure area at Iceland. Each change in the seesaw's movements has massive atmospheric repercussions. There could be above average temperatures coupled with stronger storms crossing the Atlantic resulting in a great deal of mixing, warm and wet weather in Europe and cold and dry weather in Canada. Or colder and drier air masses, weaker winds, and fewer storms across the Atlantic, and less mixing could take hold with cold weather in northern Europe and snowy conditions along the U.S. East Coast. It all depends on whether the seesaw's pressure differential is positive or negative.

A second and sometimes competing seesaw is the Arctic Oscillation which is based on pressure in the Arctic region and helps determine the position of the jet stream. When this is factored in, a phenomenon called the polar vortex comes into play, a term we heard a lot during the winter of 2015 when frigid air and mountains of snow pummeled the Northeast. Adding more complexity, our climate is affected by a third factor, originating from warm or cool ocean temperatures off Peru: El Niño or La Niña. The former produces generally dry conditions in the East, while the latter provides the East Coast with a greater chance for strong hurricanes.

These enormous climate systems are important for the Gulf Stream. The North Atlantic Oscillation is the "dominant mode of recurrent atmospheric variability over the North Atlantic," according to Kenneth Drinkwater, lead author of a study about the response of marine ecosystems to climate variability. The NAO influences weather factors: wind speed and direction, air temperatures, and precipitation. These variables lead to physical and chemical changes in the water: temperature, salinity, vertical mixing, circulation patterns, and ice formation.

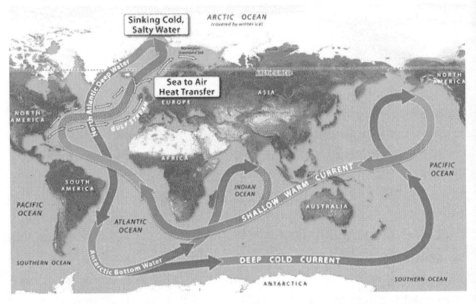

**Ocean Currents from Rahmstorf. (*Nature*, 2002)**

What keeps all of the climatic shifts going is the ocean conveyor. The sinking cold water draws more warm water up the coast to replace the water that has dropped out of the surface layer, forming a continuous circulation akin to a heart pumping blood. But what happens if one of the heart's arteries becomes blocked? The blood flow lessens or may stop entirely – with dire consequences. What happens if the oceanic pump loses strength? What would happen if it stops? We would be facing dire worldwide consequences as well. Even if it slows, there would be many consequences.

As the summer wore on, Sarah's journey did not become easier. She was going nowhere fast. The Newfoundland vortex was only the first. She rowed in big circles several times during the months. As August turned to September and then to October, she was still on the ocean. Her team constantly updated her about sea surface and weather conditions, while blog followers tracked her progress. The trip was taking much longer than she had anticipated. She had exceeded her planned margin of safety, and her supplies were running low.

Sarah was still on the Atlantic during the hurricane season. Although the mainland United States was spared from major storms, one named Joaquin found her. She had completed two-thirds of her journey, nearer to the Azores than to the United States, but still thirteen hundred miles from home. After 143 days at sea, she had to be rescued by the *Federal Oshima*, a ship heading in the opposite direction, to Montreal. She was forced to abandon both her dream of complete circumnavigation and her beloved boat *Happy Socks*.

Many people equate climate change with one consequence: rising sea levels. If projections are accurate, sea level rise is a serious concern because it would affect millions of people on the eastern coast of North America during the twenty-first century. Flooding, severe storms, erosion, and inundation would occur. The rate of sea level rise has been revised upward several times during the last couple of decades as events in the Polar Regions have accelerated. Models graphically indicate what coastlines will look like in the future. They are not optimistic renderings for many areas. Cities in the Northeast, the state of Florida, and many coastal communities in between will be affected. Cape Cod is not merely one of those places. It is in the bullseye, as is the entire Gulf of Maine where some of the greatest rates of sea level rise on the East Coast are expected to occur. And as the Cape Cod Commission has revealed, Jeremiah's Gutter would again become a waterway, leaving the Outer Cape towns from Eastham to Provincetown on an island yet again.

Sea level rise is not the only effect, though. Thomas Delworth, lead author of a report to the Congressional Subcommittee on Global Change Research, does not see a total collapse of the Gulf Stream occurring during the twenty-first century. But he identifies effects caused by even a weakened Atlantic current. His conclusions are based on recent observations and inferred from paleo signatures from the last ice age. In addition to rises in sea level due to "large-scale warming of global oceans and melting of land-based ice sheets induced by carbon dioxide," Delworth indicated an extra sea level rise of about thirty-five inches in the North Atlantic with a weakened, not collapsed, Atlantic Meridional Overturning Current that would affect the United States, Canada, and Europe. Moreover, a weakened AMOC would cool the North Atlantic, causing a southward shift of the tropical "doldrums" that would lead to dry conditions in the Caribbean, the Sahel region of Africa, and the monsoon areas of India and Asia. Furthermore, the dominant air circulation pattern in Pacific tropics would be modulated, and that could exacerbate North American drought conditions.

Climatologists, who interpret information from large sets of data accumulated over many years, do not define a trend based on one map for a single three-month period. The map illustrating a large area of very cold water southeast of Greenland after an unusually warm winter was just one dot of data, and scientists treat it as such without drawing conclusions. Granted, it persisted and was there in the winters of 2016 and 2017 as well, but that is still not long enough to make any conclusions. Each piece of information represents another dot – the video of Arctic ice melt, the data amassed over years reflecting the increasing rate of the Greenland ice sheet's decline, and observations that the strength of the Gulf Stream and the entire circulation pattern is slowing. When these dots are connected, a picture begins to emerge of vivid and evolving climate change. Meanwhile, the monitoring and measuring continue, and interpretations are refined based on the additional data.

What does it mean if the Gulf Stream and AMOC are weakening or slowing down? Overall, it means that this phenomenon is changing just about everything we have been accustomed to for generations, but we don't know what the ultimate consequences will be. This is new territory. We have pieces to the puzzle from earlier times, but we do not have a clear understanding of where we are heading.

One thing it certainly does mean is changes in the ecosystem throughout the food web. The North Atlantic is notorious for stormy seas which cause mixing which, in turn, leads to nutrient cycling that ultimately results in high productivity. It is no surprise, then, that the atmospheric conditions significantly but indirectly influence the ecology in the seas because of changes in physical and chemical characteristics. If any of the factors are "off" in any given year, it could mean cascading ecological effects.

As an example, phytoplankton production, at the very base of the food chain, is cyclical. Phytoplankton requires sunlight and nutrients to create a bloom. Winter mixing brings nutrients toward the surface of the sea where there is more light so that a spring bloom can occur. In the summer, species that thrive on lower levels of nutrients come into their own, while in the fall, phytoplankton production slows as light decreases. Zooplankton need the phytoplankton, and when their numbers mushroom, species like herring and menhaden follow the bloom, as do baleen whales. Larger fish eat the smaller fish and so it goes up the food chain, all species dependent on that initial bloom of phytoplankton in the spring. The impacts can be immense if nutrients are not available for phytoplankton because of a change in upwelling. This whole process depends on the oceanic conveyor. As it changes, so may the phytoplankton production cycles.

Adapting to a new paradigm of ocean resources is a necessary part of the ecological equation. Many species exist within fairly narrow temperature ranges. Some species are already moving north, on land and at sea, as their environment shifts. Empty niches will be filled, but predicting how all of the shifting ecosystems will be sorted out is unknown. Predator and prey relationships, migratory patterns, and plankton communities that are the base of the food chain are all critically important. If species adapt at different rates, existing species relationships may be significantly altered. Species that have never been preyed upon by certain other species, or species that can no longer find their natural food source, may or may not survive because all of them jockey for position within the ecosystem. Species may be threatened by diseases that have never threatened them before or that they have been able to control.

Opportunistic species may overcome traditional ones. In short, the marine ecosystem, both in estuaries and at sea, may be completely altered from what we now know.

This uncertainty means that our outlook on harvesting creatures from the sea will have to adapt as well. Fishermen will have to seek out different species or adjust their methods, or both, as they have done in the past; and shellfish aquaculturists will have to modify their culture methods to meet the new challenges.

We are facing many facets of climate change and the uncertainty that goes with it: a global heating trend; melting polar ice caps; a deceleration of the engine that drives the global oceanic conveyor; the climate in general and our weather in particular; paradoxical consequences due to extremes in weather; tremendous ecological adjustments; and unprecedented sea level rise. Our future and how and where we live will change profoundly. There will be shifts in ecology, in economics, in social conditions, in resource allocation and management, and in the world we leave for future generations.

Sarah Outen received unexpected news on January 5, 2016, three months after she had returned to the United Kingdom, from the Royal National Lifeboat Institution in Ireland. They had spotted a small boat that appeared to be in trouble about a half mile offshore from Valentia and sent a boat to investigate. They had found *Happy Socks*. It was a lucky day because Sarah's boat was on course to be smashed against the unforgiving rocky coast. Encrusted with barnacles and draped with seaweed, the little boat had been on the Atlantic for some 230 days.

The Gulf Stream was still working, although slower than expected. *Happy Socks* had survived her voyage intact and made it home. As had her skipper.

*Happy Socks* **comes home. Sarah with Royal National Lifeboat Institution in Ireland. (Courtesy of Sarah Outen**)

Chapter 23
# Challenge and Inspiration

The storm forecast for October 30, 1991, was described as either an unusual late-season hurricane or an early-season nor'easter. Regardless, people scrambled to prepare. Boats still in the water were hauled or moved to safe, protected locations. As the storm approached, it was obvious that this was going to be a doozy. The worst of it would hit at high tide in mid to late afternoon. Normally, high tide occurs about three hours earlier in Cape Cod Bay than in either Nauset or Pleasant Bay, but a big storm changes that. The wind pushes the water landward, and there is no real low tide. So as the tide gets higher and higher, surging farther and farther inland, the time of high tide becomes about the same everywhere.

"Storm watch," as I called it, was an official part of my job, and I had permission to go anywhere. First on the list was Rock Harbor on the Cape Cod Bay side. I couldn't get to the harbor. Water flooded the road. And the tide was still coming in. I checked a few other bayside locations and took photos and then headed for Nauset. The tide was supposed to be high at around 4:30 p.m., but at that time of year, late October, it got dark early, and I had hoped to see the tide from all over town. Irritated, I couldn't see the worst of it in the fading light, but what I could see was incredible. As it began to get dark, I tried one more trip to Pleasant Bay, but I couldn't get near the water. Route 28 was closed. Waves were crashing over the road.

The next day, the wrack lines of seaweed that had floated up on shore with the tide told the story. They were the highest in anyone's memory. Boats looked like toys thrown up into waterfront yards. Huge square-edged logs, usually seen in parking lots or marinas but not normally along the bay shores, were scattered as well. Several feet of water still covered the parking lot of the shopping center, across from Town Cove, that had been built on a cranberry bog and was part of the Jeremiah's Gutter drainage system. High-water lines marked the sides of buildings like the Orleans Inn, built in the nineteenth century by the Snow family where packet ships brought supplies for their burgeoning business, still in operation. Its ground floor was flooded by the storm.

I had to wait a couple more days for a low tide before seeing what had happened to the coastal banks. Stairways to the beach were gone or mangled and scattered. Shrubs and grass had disappeared, and trees were uprooted and lying on the beach. The banks were sheer cliffs of sand and clay. Long-buried Native American shell middens, covered by centuries of accumulated soil, were exposed once again. Houses at the top of the bank were closer to the edge than they had been a couple of days earlier.

Later on, Sebastian Junger gained fame when he called it *"The Perfect Storm,"* the title of his book. Warm air from the remnants of Hurricane Grace had merged with cold air from an offshore storm, forming a ferocious tempest that backed into New England. Its ferocity made it a hundred-year storm along the ocean side of the Cape.

Storms that cause flooding or erosion or both are only one of the challenges facing coastal regions like Cape Cod. There are others: climate change that ramps up uncertainty about the biological and physical aspects of resource management; achieving consensus for resolving environmental, financial, political, and social issues involving wastewater and degraded water quality; changing traditional management approaches to encompass ecosystem-based resource management; reconciling constantly-shifting baselines from a population always in flux because of a high rate of resident turnovers; and developing multimedia strategies to keep that population informed.

Storms with the intensity of the one in 1991 are projected to occur more frequently than every hundred years because of climate change. Flooding is a consequence of storm intensity, and Cape Cod is particularly vulnerable because of its vast low-lying areas. Flood maps issued by the

Federal Emergency Management Agency and updated every decade illustrate the magnitude of the Cape's problem. The maps show velocity zones that would be affected by waves and flood zones susceptible to rising water. They depict the zones vulnerable to a dangerous hundred-year flood and a catastrophic five hundred-year flood. Much of the area has been filled in or developed. Clearly, however, Cape Cod faces the kind of trouble that no amount of engineering can prevent because the Cape, as we know it early into the twenty-first century, is disappearing into the sea.

Undeveloped barrier beaches physically change in shape and locations. We watch in awe as the power unleashed by particularly nasty storms rearranges sand dunes, rips out beach grass, and carves new conduits for water to flow in and out of the bays, and we no longer know what is normal. That movement is clearly evident over a relatively short time – one winter, a year, or several years. Chatham Light provides a particularly vivid public overlook to gauge the change. Old-fashioned viewfinders in the parking lot provide fixed points to view various sand bar configurations in the harbor.

Storms like the one in 1991 cause wash-overs that may cut new inlets through the barrier beach. The Chatham breaches in 1987 and 2007 and the Eastham breach in 1993 became navigable inlets and remind us that the Cape is geologically active. The land behind the dunes is protected because the beach bears the brunt of the storm. Because Nauset Beach is not developed, it behaves naturally, continually providing that protection.

Conversely, when barrier beaches are developed, covered with roads, houses, hotels, restaurants, gift shops, and other amenities, the beach cannot move. All of the buildings on the beach are endangered and can be easily destroyed. Newscasters frequently use such locations to show a storm in progress.

Where there is no barrier beach providing natural protection, the storm hits the land directly, causing substantial damage to homes and businesses facing the ocean. Chatham experienced just such a scenario when a new inlet formed in 1987. Much of the barrier beach eroded and shifted, leaving a wide gap. Landside homes suddenly faced the open ocean without any protection. In the summer, it was beautiful and placid. But when winter arrived and storms hit, houses were endangered and several of them fell into the harbor.

Dealing with erosion and flooding are difficult, but the bottom line is that those conditions will persist and probably get worse. Planning for the effects of climate change – the likelihood of flooding and erosion – is essential. Cape Cod is built on sand and glacial till which is neither stable nor permanent. How Cape Codders accept that fact will determine what the threatened areas look like in future decades. Developing traits such as resilience and adaptation to deal with the changes are important for the future. Yet human nature resists that notion.

The threat of debilitating storms looms large. The sand moves, taken from one area and deposited in another. Erosion and deposition are nothing new. They have been occurring since the glaciers retreated. But some areas are experiencing measurably larger movements. Climate change ups the ante with fierce storms that challenge the lives of those who live closest to the sea. Erosion, many feet per year in some areas, imperils houses and other structures atop coastal banks. Some structures can be moved back to safer places. From 1996 to 2015, five iconic lighthouses, including Highland Light in Truro, Nauset in Eastham, Sankaty on Nantucket, and Gay Head on Martha's Vineyard, were moved hundreds of feet inland to prevent them from falling into the sea. Other structures that can't be moved will inevitably tumble over the edge.

Cape Cod and the Islands are not alone. Cape Hatteras light in North Carolina joined the list of lighthouses moved farther from the encroaching sea.

But it is not just lighthouses that are affected. Communities and cities all along the Atlantic coast, such as Boston, much of which was built on filled marshland, are struggling with how to plan for and deal with sea level rise because of climate change. The *Boston Globe* unveiled a plan in February 2017 to construct a storm barrier from Deer Island to Hull, a distance of four miles. The sobering article described the multibillion-dollar project as among the largest of its kind but on a par with barriers planned or completed in New Orleans, Venice, and Rotterdam. Boston's barrier would be twenty feet above harbor waters at low tide. It would be constructed with gaps wide enough for shipping but equipped with gates that would close those gaps for storms, and would end up being a barrier that would reduce both the water flow from regular tides and the ferocity of waves during storms, wrote David Abel.

Without the barrier, planners estimate $1.4 billion per year in property damage from flooding if seas rise by three feet above the levels of the year 2000, but new estimates by the National Oceanic and Atmospheric Administration suggest that without curtailing greenhouse gases, seas could rise by 8.2 feet by 2100 instead of the 6.6 feet they had estimated earlier. New York and New Jersey are similarly endangered, as we saw when those states experienced devastating damage from Hurricane Sandy in 2012. And map projections for the Florida peninsula show an impressive amount of the state that will be flooded.

As the agency most associated with both climate and marine resources, NOAA understands threats to communities in a rapidly-changing environment. The agency has developed interactive tools to help communities understand the projections in order to foster meaningful discussions, plan accordingly, and adopt adaptive measures. On June 15, 2016, NOAA announced an $8.5 million grant program called Coastal Ecosystem Resiliency, with awards of $100,000 to $2 million to begin the process. The grants will fund programs that protect life and property, safeguard people and infrastructure, strengthen the economy, or conserve and restore coastal and marine resources.

Climate change is also altering established assumptions about populations of countless species in the North Atlantic. The longstanding biogeographic dividing line, where Cape Cod marks the southern boundary of northern species and the northern boundary of southern ones, may be shifting, according to a 2010 Environmental Protection Agency study, by Stephen Hale, of inshore coastal areas from Delaware Bay north to Passamaquoddy Bay. Climate change and warming seas appear to be creating what Hale calls a "leaky" line, rather than a solid one, as species begin to cross it.

As examples, blue crabs have moved slightly north and are now found in Nauset and in Cape Cod Bay communities farther north. Mantis shrimp that were formerly absent have also been found in Nauset, while green crabs are now as much of a nuisance predator in Pleasant Bay as they are in Nauset.

Lobsters offer a larger perspective. They used to be prevalent from Long Island Sound northward. Now, lobsters are in short supply in Long Island Sound, and those that are there reveal signs of stress such as shell diseases. Lobstermen in Nauset and Chatham have to set their pots farther north than they did just a decade ago. Hook and net fishermen are also finding fewer numbers of the species they used to catch, like cod, and more warm-water species.

If the changes we have already witnessed are any indication, there will be multiple repercussions as water temperatures rise with climate changes. Predator and prey relationships, species abundance, diseases, fouling, growth, and unexplained mortalities are likely ecological alterations. Documenting the changes and understanding their implications is yet another challenge. It is one reason that monitoring is especially critical now. The environment, especially everything affiliated with the biogeographical zones from the Gulf of Maine to Long Island Sound, is changing so rapidly that it is difficult to keep up, let alone try to grasp the significance of all the changes.

The Gulf of Maine is warming faster than ninety-nine percent of the planet. The reason is not clear yet but changes in the currents undoubtedly play a large part. Everything we now know about the marine environment of the area is likely to be different in the foreseeable future. It is like throwing a deck of cards into the air and guessing the order of how they will land. We can assume the right whales that showed up in the Gulf of Saint Lawrence in the summer of 2017 were following the copepods they feed on. Does that mean that the food moved north from the Bay of Fundy to the Gulf of Saint Lawrence? Was that summer event an anomaly or a new norm? The answer is not known. But it is only one example of complex interactions and temperature changes.

For shellfish, there is cause for concern brewing. The carbon dioxide in the atmosphere related to climate change presents a worrisome chemical

269

change – ocean acidification. Briefly, this means that the excess atmospheric carbon dioxide is dissolving into the seas. As it mixes, it creates carbonic acid that can disintegrate shells. That means that any animal with a shell – including plankton, the building blocks in the seas – are potentially vulnerable.

A few days after fertilization, shellfish larvae develop a thin shell. Then, as swimming larvae, they form their hard shells after about two weeks. They cannot survive if their shells cannot form because of acidification. Hatcheries on the West Coast have been plagued with this problem for several years and now routinely buffer the incoming water to decrease its acidity. Hatcheries on the East Coast also pay special attention to the acidity of their incoming waters, and at least one hatchery in New England reports acidity issues. We do not know how ocean acidification might affect shellfish on an estuarine level, but research continues. Meanwhile, ocean acidification, its effect on animals with shells, and the uncertainty of what will happen as the climate continues to change is another reason to exercise caution when debating whether to rely on shellfish as a long-term wastewater management solution.

Shellfish growers are acutely aware of the problems associated with climate change on their farms. In 2018, seven prominent East and West Coast growers partnered with the Nature Conservancy to launch the Shellfish Growers Climate Coalition. They aim to "shine a light on the many ways climate change already is affecting food production in the United States," their press release states.

The magnitude of challenges for life in the seas that ocean acidification alone may present is mindboggling. How ocean acidification will affect the entire food web, from plankton at the base all the way up the chain, is unknown. And that means that everything that lives in the estuaries, Gulf of Maine and ocean beyond can be affected, a thought that is difficult to contemplate. Research is being conducted in the United States and Canada to better understand the reactions of various marine invertebrates to increased levels of carbon dioxide with its higher acidity and lower amounts of available carbonate for shell production.

As we watch events unfold, some quickly and others over time, we will be faced with a radically shifting baseline for numerous species simultaneously.

In the shellfish world, there is cause for hope in spite of challenges. People are beginning to think of marine aquaculture in terms of its benefits – a green industry that produces high-quality protein. And it produces that protein by filtering natural phytoplankton in the water, unlike sources of land protein including chicken, beef, and pork, that use fishmeal among other sources of feed. The industry also provides jobs and societally beneficial ecosystem services. Growers are beginning to integrate edible seaweeds on the farm with shellfish using the entire water column. Some researchers are adding a third component, fish, to the farm.

The integrative approach addresses acidification as well as nutrient enrichment. It is not a new concept. The Chinese adopted it many centuries ago. The modern version is small in scale at this point, but companies are investigating its large-scale possibilities.

Resource management based on an ecosystem-wide approach is another challenge. Shellfish managers work with species that do not move very far, if at all. That makes them easier to deal with than species that swim freely, but managers must still take many factors into account. Shellfish managers try to gauge the amount of shellfish in the water, whether it is mussels that are usually visible on intertidal flats or clams or quahaugs that live beneath the surface, in order to set limits. They look for predators to check for mortality rates and evaluate the annual amount of new juvenile seed. They observe changes in the environment, like sand covering a productive clam bar, or fouling species that may smother shellfish, or a buildup of seaweeds along the shore that may do the same thing.

When the managers set harvest limits, they are making assumptions. One is that licensed individuals will not always harvest the maximum amount they are legally allowed. But what if everyone who had a license decided to take their limit of say, mussels, every week that they could? How long would the population last before the amount taken exceeds the ability of the mussels to regenerate?

Shellfish "stay put." But consider the difficulties of managing populations of fish that swim throughout their range. What happens to species that take years longer than shellfish to mature? And what happens if you underestimate how many there are to take in the first place and set a quota above a sustainable level? Then it becomes clear how any resource can be overharvested when only a part of the problem is in focus rather than the whole picture. What about whales and seals that are illegal to harvest but that fall under the umbrella of the managing agency that also manages fisheries. What if the mammals compete with us for fish that we do harvest? How do you deal with all of them when they interact with other animals and with us within the same concepts? These are not easy questions to answer, but ecosystem-based management looks at the interactions between and among other species as well as man's place in the scheme of things to formulate options.

Resource management plans employ the ecosystem-based approach. Shellfish was one of the major resources considered for a comprehensive resource-based management plan for Pleasant Bay, developed on a bay-wide basis. The plan was approved in 1998 by the four towns around Pleasant Bay — Orleans, Chatham, Harwich, and Brewster. It focused on five key management topics: biodiversity, boating safety and navigation, public access, shellfish and aquaculture, and structures such as private docks and revetments. Thanks to the participation by hundreds of stakeholders, municipal employees representing various interests, and volunteers, workgroups identified and mapped natural and human-use resources. Using Geographic Information Systems technology, maps overlaying one resource on top of another were created. Those maps gave people a better sense of where the resources were and what might impact them, essential information for forming the final plan.

Ramping up ecosystem-based management from a mussel flat or a watershed or even Pleasant Bay to larger ecosystems such as Stellwagen Bank and the expansive Gulf of Maine is another challenge because of the greater complexity. The Stellwagen Bank National Marine Sanctuary's 2010 management plan exemplifies how these programs can be broken down into smaller manageable elements.

NOAA, which administers the marine sanctuary, based the

management plan on 670 source documents, most of which were peer-reviewed published scientific papers. Those studies were not necessarily confined to specific sanctuaries but were the results of surveys of a much broader geographic region within the Gulf of Maine. The process included a human-use component that was much broader than the same concept for Pleasant Bay, given the size and use of the gulf. It included three hundred people participating in scoping sessions, two hundred serving on issue-based working groups, and comments from twenty thousand citizens. What emerged from this effort was a picture of the natural and human resources that are important to the sanctuary.

Developing a management plan for the Stellwagen Bank sanctuary, in the middle of the Gulf of Maine, with its incredible array of marine resources, has been an immensely complex challenge. With today's technology – GIS, three-dimensional graphic imaging, computers capable of analyzing large sets of data, and showing data points graphically – pictures emerge that help formulate management plans.

One example of the value of this type of mapping was that shipping

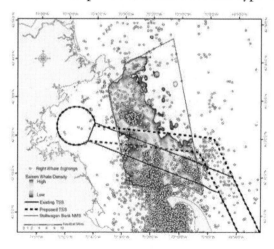

lanes were identified as a problem because of the potential of ships and whales colliding. The Stellwagen Bank sanctuary has the most commercial shipping traffic in the Gulf of Maine. Planners created a picture of the spatial distribution of whales within the sanctuary from 1979-2004. The data

**Stellwagen Bank shipping lane alteration for route to Boston. Circles represent whale sightings. Solid line indicates former route through high number of whale sightings. The dashed line represents the altered route through the area with fewer whale sightings. (Stellwagen National Marine Sanctuary)**

273

represented 255,000 observations by whale watching vessels over twenty-five years and showed two distinct areas of heavy whale activity with a relatively low activity area between them. The peaks coincided with sandy bottoms preferred by sand lance and herring – which whales eat – while the less-used area had a muddier bottom. When a map of shipping lanes to Boston was overlaid on top of the three-dimensional map, it revealed a substantial overlap of ships and whales. But it also provided the possibility of a compromise that would reduce the likelihood of ship strikes if the lanes were shifted to the area of lower whale sightings. Researchers projected an 80 percent reduction in ship strikes when analyzing data about whales within the old lanes and the proposed new ones. After considerable discussion with stakeholders, including those who were involved with maritime shipping and commerce, the lanes were shifted in 2007, a difficult and costly decision for the shipping industry.

Completely preventing ships from striking whales is impractical considering the volume of shipping. Moreover, it is virtually impossible to determine the actual reduction of ship strikes and to gauge if the shift is working because whales swim at will in and around the shipping lanes. But the compromise exemplifies the possibility of positive results with ecosystem-based management.

Similar to the Pleasant Bay plan, the Stellwagen Bank plan, authorized by the National Marine Sanctuaries Act, relies on analysis of long-term data about a multitude of topics, continuing research, and monitoring, all of which takes financial resources and time. That type of synthesis is often considered unimportant or too expensive by potential funders who do not project an immediate benefit. Funding is regularly granted for projects lasting up to two years, which is problematic given the reliance on long-term data. Piecing together data from many different studies requires researchers to spend time analyzing the information, increasing the cost of the plan. Although continued monitoring or repeating field investigations are frequent recommendations at the end of a project, such efforts are rarely funded. Instead, new projects are most often given higher funding priority, leaving monitoring on a back burner.

Moreover, the process of ecosystem-based management, a concept that deals with enormous variables and complexities, is frequently castigated by those who want simple solutions to complicated problems.

Regulators, charged with administering laws designed to protect environmental interests, have a difficult role, especially when the laws are based on pieces of the puzzle and not the whole ecosystem. NOAA's National Marine Fisheries Service is an example of an agency that has a tough job. It is responsible for "stewardship" of the nation's living marine resources, their habitat, interactions, and ecosystems. The agency's directive is derived from five key statutes: Magnuson-Stevens Fishery Conservation and Management Act; Endangered Species Act; Marine Mammal Protection Act; National Aquaculture Act; and National Environmental Protection Act. The agency must address all five mandates simultaneously and in an ecosystem-based context. Facets of ecosystem considerations are stated in each of the acts, but an overarching, encompassing ecosystem-based approach is needed to achieve the objectives.

Fisheries management has functioned on a species-by-species basis for decades. However, in order to meet its mandates, NOAA is changing how it operates – to ecosystem-based management. The concept is not new. But NOAA's decision to embrace it is very new. NOAA released a draft of its Ecosystem-based Fishery Management Policy in the fall of 2015. What the final policy will look like or how it will be implemented has not been determined. But it is an ambitious undertaking and underscores the necessity to consider both Mother Nature and human nature for ecosystem health and vibrancy.

Managers will not be dealing with individual fish species populations in a vacuum, since population dynamics is only one aspect. Additional aspects such as interactions with other species, environmental changes and their effects, and pollution and other stresses on habitats and water quality will also be addressed. However, it takes a vast amount of research, monitoring, collaboration, and cooperation to initiate ecosystem-based management, as the example of the shipping-lane change indicates. And that means funding in a period of fiscal constraint.

There is another challenge as fisheries and mammals swim across international borders. The incident report following the summer mortalities of North Atlantic right whales in the Gulf of Saint Lawrence clearly states that there is a gap in the knowledge about the use of habitat by whales in Canadian waters. Data about where whales are concentrated, research into reducing entanglements with fishing gear, and analyzing the conflict between shipping and whales are all needed before forming management options. The necessary discussions in Canada will most likely be difficult in the face of a rapidly changing environment and the decreasing North Atlantic right whale population.

**Volunteers build an oyster shell reef as part of South Carolina Oyster and Restoration Enhancement program. As of 2014, more than twenty-five thousand volunteers have used more than 1,100 tons of shell to build 225 reefs at 69 sites. (Courtesy South Carolina Oyster and Restoration Enhancement Program)**

While many people see a rather depressing picture of where we are heading, some remain hopeful about the future.

In many areas of the United States, including Cape Cod, groups of volunteers are supporting shellfish restoration projects such as building oyster reefs, collecting shells, cleaning upwellers, and planting shells as cultch – a substrate that setting oysters attach themselves to. These folks smile when they get wet and dirty and enjoy a sense of "giving back" to the community. The volunteers learn about the estuaries, water quality, predator and prey relationships, fouling organisms, and many other facets of shellfish management through their work. Then they spread the word to other residents. Other groups lend a hand as "citizen scientists," some collecting water samples and others participating in beach cleanup days. The educational benefits of these activities alone is incalculable. The number of volunteer hours can be staggering, and that information helps to convince funding agencies that there is community support for the programs. School groups are instilling an appreciation of the marine world, like the Billion Oyster Project in New York City or the Sound School in Connecticut or South Carolina Oyster Restoration and Enhancement Program. Civic-minded people, often retirees, volunteer to participate in programs like the Suffolk Program in Aquaculture Training (SPAT) in Southhold, Long Island, or ReClam the Bay in Barnegat Bay, New Jersey, or oyster gardening projects, or Save the Bay programs on all the coasts. All participants are learning about the bays in their own back yards and work to improve water quality and the overall health of estuaries.

Hope for the future on a grand scale came in December 2015 when 196 countries agreed to an unprecedented and historic pact during the United Nations Climate Change Conference in Paris. Encouraged by small island countries in danger of disappearing into the sea and by developing countries, the pope and nations large and small charted an ambitious course for reducing carbon dioxide emissions to preindustrial levels.

A long-term commitment will be necessary for the climate pact to be effective, but it is an essential beginning. Christine Figueres, executive secretary of the United Nations Framework Convention on Climate Change, put it this way: "A critical mass of people in every sector, in every geography, at every age, with every conviction decided to rise to the higher purpose."

And Cape Cod was represented. Richard Delaney, executive director and CEO of the Center for Coastal Studies, participated in the conference, sharing success stories and the needs and concerns of Cape Cod with those in the global community.

On June 8, 2016, World Oceans Day, the Center for Coastal Studies and Pleasant Bay Community Boating cosponsored a symposium called "From Paris to Pleasant Bay." Delaney was the principal speaker. His optimism for the future was reflected in a handout for those who attended.

"Climate change. The words evoke something far away – in time, in space. Polar bears and caving ice sheets. Droughts in Africa; monsoons in

Asia. Islands underwater in 50 years. You know it's real. You know it's scary. But it's too big to think about, too overwhelming to envision how we can make a difference. Which makes it all too easy to put it out of our minds for another day.

"But it's not going away. And each of us can make a difference. In the same way every ocean, every small bay and estuary is made of billions of drops of water, every large problem can be chipped away by small actions. What each of us does every day, in our small piece of the world, makes a difference."

Delaney then announced the creation of a new entity: the Cape Cod Climate Change Collaborative, the "5-Cs." The collaborative is "a regional campaign to unite the varied expertise and experience of Cape Cod organizations to address the impacts of climate change," he explained. "Here on the Cape we have an acute understanding of the threats of climate change – warmer ocean waters, sea level rise, and increased storm intensity. The goal of the Collaborative is to spur individual, local, and regional actions to mitigate these impacts by linking available knowledge, talent, and tools of Cape Cod organizations."

Collaborative partners at the inaugural event included environmental groups; municipal, county, and state representatives; and economic development interests. Each organization explained who they are, what they do, and how that relates to mitigating climate change. Projected programs, how to join in the effort, and additional things individuals can do were also discussed. Many of the organizations have other partners, and it is likely that the collaborative will grow.

The following Saturday, June 11, the Center for Coastal Studies celebrated its fortieth anniversary. Rich Delaney told the 350 people attending the gala event at Provincetown Town Hall that he looked forward to having 196 organizations join the climate change collaborative, just as 196 countries had signed the climate change accord in Paris. The collaborative is off to a good start toward achieving that goal.

As we look back at what we have done to bring marine resources to this point, we also have an opportunity to look ahead and, after acknowledging our collective mistakes, create a new approach.

It is clear that the far north is experiencing the most dramatic changes, but the Northeast is in the forefront of projected accelerated changes. The Cape Cod Climate Change Collaborative could serve as a model for other communities and for larger areas such as the Gulf of Maine and Chesapeake Bay, or any region that embraces the idea to move forward.

It's the beginning of a new joint venture. Cape Cod, that small, seemingly insignificant arm of sand sticking out into the Atlantic, is providing the leadership and a model to address the enormous challenges of climate change by focusing attention on this region – by thinking globally and acting locally.

Chapter 24
# Tide Together

Herring Cove Beach in Provincetown is a vantage point that symbolizes the complexity of the marine environment. The roles of mammals, turtles, fish, shellfish, and thousands of other species; the watery environment they call home; and the ways they have interacted with us all come into focus from this beach.

In thinking about the ecosystem as a unit, a biological axiom comes to mind: You need adults to have babies and babies to have adults. It doesn't matter if they are humpbacks or humans, cod or clams. From an ecosystem perspective, there is a corollary. You must acknowledge the pieces to see the whole, but you must have a picture of the whole to manage the pieces. One doesn't work without the other. We have spent centuries dealing only with the pieces that affect us. And it has taken us centuries to realize that it doesn't work.

At Fort Hill, the beauty of the estuary, the marsh, the barrier beach, and the ocean beyond are sharply focused. My experience taught me to see beyond the beauty to the challenges and to our relationship with marine resources begun in the past and projected into the future. From a different vantage point, I again marveled at the beauty but saw what lies beneath.

A Dolphin Fleet Whale Watch boat traveled north past Herring Cove Beach for a whale-watching cruise on a bright fall day. It was headed toward Stellwagen Bank where humpbacks had been spotted. Maybe one would breach, leaping out of the sea in a gravity-defying maneuver. Or maybe, in the distance, passengers would see a sleek fin whale skimming the surface. It was the wrong season to see a V-shaped spout of a highly-endangered right whale still in the bay where half of the world's entire population comes in the winter and early spring to feed.

A commercial fishing boat passed offshore of the whale watch boat, heading back to the harbor. It was one of the few remaining finfish boats at MacMillan Wharf, a far cry from the large number of boats that called Provincetown home several decades ago. The captain's day would not be over when he tied up at the dock. He would have one more task: notifying the National Marine Fisheries Service about what he had caught, where, and how much.

Across the bay, the hills of Plymouth, the second place the Pilgrims had landed – after leaving Provincetown's protected harbor – and the site of the first European colony in Massachusetts, were clearly visible twenty-five miles away. The Pilgrims had considered Provincetown an unwelcoming place to settle. The apparent lack of fresh water and the sand dunes were serious impediments to them. They marveled at the fish and whales they saw, but the Pilgrims had a European way of looking at natural resources as something to exploit. That was vastly different from the Wampanoags' attitude about living in harmony with those same resources that they believed to be gifts from the Creator.

The asphalt parking area at the beach was just wide enough for cars to pass one another and to park facing the sea. It seemed squeezed in between the sandy beach and the unwelcoming dunes the Pilgrims had seen that stretched across the land from Cape Cod Bay to the Atlantic. National Seashore maintenance crews scraped the parking lot every spring to keep it open from the drifting sand that tried to encroach and obliterate it on the east side and from the erosion that threatened it on the west. Now, the National Seashore has given up. It plans to move the parking area farther inland rather than continually fight the inevitably encroaching sea.

In between Herring Cove Beach and Plymouth lies the curved Cape Cod, the hook-shaped sandbar extending over forty miles out to sea from the mainland. Not visible were the numerous harbors, marsh creeks, and estuaries in every town that are a vital link from land to sea. Also not visible were the vast intertidal and shallow areas of the bay where cages and bags are filled with oysters and where wide swaths of netting cover quahaugs, the tools of the trade for shellfish aquaculture. Other industries had come and gone. Whale huts and blubber-rendering try pots as well as fish flakes and salt works were absent from the shore that was now lined with homes offering magnificent views of the bay. But every house and business built on the shore or farther inland produces nutrients and toxic substances that flow into the bay. All of these factors are part of the ecosystem of Cape Cod Bay.

But looking at the expansive bay in its entirety, it represents the picture on the cover of a jigsaw puzzle where the beach, whale watch boat, fishing boat, estuaries, shellfish aquaculture, houses and business on the land, and the entire marine world beneath the surface of the bay are all individual pieces of the puzzle.

Clearly, our collective procedures of taking whatever we want from the sea in any size or amount and whenever we want it have not been sustainable. It makes no difference which marine resources are being discussed. We have nearly exterminated whales, seals, otters, and other marine mammals. We have indiscriminately harvested fish of all sizes and species, causing fisheries to collapse around the world. We have harvested so many oysters that they have become functionally extinct.

Fishing communities have morphed into tourist centers where working waterfronts are wedged among businesses that have no direct relation to the water or have simply disappeared along with the fishermen. Meanwhile, what we do on the land also affects the seas. Estuaries are slowly choking to death from residential and commercial wastewater that is adding excess nutrients to the groundwater, destroying habitats for fish and shellfish in the process. Other land-based activities, including those on a very large scale, are causing oceans to become more acidic from carbon dioxide, threatening the very building blocks of oceanic life.

Adhering to existing laws has been problematic. Species-by-species management that does not account for interactions among all species within the same ecosystem has obviously not worked. We need to change our approach to marine resource management. We are beginning to do that – finally.

Our penchant is to regard marine resources in solely economic terms of how they benefit us, species by species. That approach is usually at the obvious expense of the exploited species, but it also often wreaks havoc with other parts of the food web, as the situation involving sea otters revealed.

Despite this, there have been some bright spots – with whales, with fish, and with shellfish. Commercial whaling has evolved into whale watching, a multibillion-dollar industry that began on the East Coast in Provincetown. And to help save entangled whales, dedicated people risk their necks to free the animals from rope shackles and then teach people around the world the skills they have perfected. Stellwagen Bank National Marine Sanctuary, where whales congregate, is itself a shining example of the area's fabulously rich and diverse environment which also epitomizes the conflicts between the oceans' resources and our wants and needs. Resilient and adaptable fishermen have banded together to express their concerns about the region's industry and now work with regulators to benefit fish and all fishermen. Volunteers give unselfishly of themselves to save cold-stunned turtles and stranded marine mammals, to help towns raise shellfish for the public fishery, to monitor herring and eel populations, and to take water samples as citizen scientists. Private shellfish aquaculture generates jobs on the water and food for the table and provides vital ecosystem benefits such as water filtration and habitat enhancement. And municipalities continue to grow shellfish for public waters. Some Cape communities have taken baby steps while others have taken giant leaps forward with wastewater management; indicating that they are acknowledging the problem. And, at last, NOAA is developing policies for ecosystem-based management, the brightest star yet. Individually, these signs are pinpricks of light, but combined, they form a potential constellation of change for the future.

For centuries, the seas, bays, and estuaries have given us gifts – food, jobs, economic rewards, ways of life, and community structure. We took those gifts for granted without any expressions of gratitude.

After all we have taken from the sea, fulfilling our desires to have it all while paying little heed to the needs of the creatures we have exploited, what gifts can we now give back? How can we correct the insults to the habitats that we have altered?

At the inaugural meeting of the Cape Cod Climate Change Collaborative, member organizations recommended things that people can do: form, join, or support local groups dedicated to protecting local resources in ponds, bays, and harbors; vote for candidates who have environmental visions; attend programs designed to educate the public about these ecological issues; care for your own property in an environmentally-friendly manner and consider how your actions will impact your entire region; and conserve water, power, and resources.

These are part of a three-step process for dealing with the issue.

The first step is understanding: to see our own actions as part of a larger picture and not isolated from the interconnected nature of things.

The second step is mindful action: to recognize that even seemingly small actions have consequences, positive or negative, subtle or forceful, singly or cumulatively; to remember that "cide," as in herbicide, pesticide, and biocide, means "killer of;" to observe fishing and shellfishing regulations; to avoid fuel leaks and keep absorbent material onboard; to pay attention to the creatures in the sea.

The third step is positive action: to volunteer to count herring or look for turtles or clean a shellfish upweller or conduct a beach cleanup; to join a group that advocates for natural resources; to practice conservation measures; to fish and boat responsibly and do whatever you can to prevent further harm to the environment.

Collectively, we all can make a difference, no matter how small each step may seem.

We still have swirling currents, and we may always have them. But our very existence depends on whether we can learn from the past and move forward to a more balanced future, interacting compassionately with the ocean at our doorstep.

# Epilogue

No cars are parked at Fort Hill, overlooking the Cape Cod National Seashore, on this bitterly cold and windy day in late January. Winter grips the land as the natural world predominates, not the one where the people with their cravings try to control it. A couple of inches of two-day old snow blanket the ground, and tufts of khaki-hued grass poke through the white that covers the field beside the road that leads to the top of this high hill.

Mountainous ocean waves are visible from up here, a reminder that Mother Nature is flexing her muscles to show who is really in charge. It is a pretty simple task for her whenever she so chooses. The fawn-colored marsh reveals small sparse patches of snow as does the high outer-beach dunes north of the inlet, although the lower dunes to the south are brown and bare where wash-over cuts are clearly visible, a result of harsh northeast storms. Although the beach is scarred, it is not broken. Yet, there is no doubt the beach has migrated farther north than ever before.

The trees at the bottom of the path from the Fort Hill parking area blocked the view of the inlet two years ago. Now the inlet has moved north of the trees, and it is clearly visible. At some point, ocean waves will wash over the dunes farther south creating a passageway for water to flow to the other side. And successive tides will continue to wash through, deepening a narrow depression. Then the outgoing tide will gush from the bay to the ocean, first as a small stream and within a few days, there will be a new inlet, the beach reacting to natural conditions as it has always done. The old inlet will fill in, and the beach will begin its northward march all over again.

A hawk, never beating its wings as it glides aloft, banks to the left and drops out of sight near the water's edge, while a single gull adjusts its flight with a few more vigorous wing beats. A flock of small ducks cruises low over the water and lands near the marsh's bank on the Town Cove side of the hill in the lee of the stiff northwest wind. Even birds are few, compared to other seasons, on this blustery day. Nature rests during the winter when the hours of darkness exceed those of daylight. Some Native Americans call it a time of "rest and cleansing," a revitalization to save strength before the land comes alive again in the spring.

A few frigid weeks and plunging temperatures later and in another part of town, there are no cars at Portanimicut Landing. No boats moored in the mostly frozen Little Pleasant Bay, none tied to the bulkhead. The only open water is about two hundred feet of the creek at the bulkhead and a narrow, shallow fringe along the shore where hundreds of waterfowl — blacks, mallards, buffleheads, scaup, geese, brant, and other birds congregate and forage in the open water, in the lee of the bracing northwest wind. A group of about thirty ducks and geese, heads underwater and tails up, search for food in the frigid creek that flows freely to the bay, unconstrained by ice that fills Pau Wah Pond up the creek from the bulkhead. It is idyllic. The land across the creek was purchased by the town decades ago to protect it from development and remains natural, with walking trails and vantage points offering spectacular views.

Across the bay, Sampson and Hog Islands also remain in a natural state, land that belongs to a family trust, now an alliance with the Cape Cod National Seashore so that the islands will never be developed. In the distance, the great barrier beach protects the bay from the vast ocean beyond.

A few scallop shells are scattered on the blacktopped parking area. The scallops, freshly killed by water birds, with scraps of the adductor muscle still attached are a welcome sight to those who expect them to be gone from this place.

In the winter, it is difficult to imagine that this gorgeous bay is endangered, something that would have been inconceivable to those who came before us. The ducks and geese, going about the business of life, indicate the continuity of nature, and the scene could well have been the scene in the early seventeenth century except that the barrier beach would have been farther away.

One wonders if the same scene will greet people several generations from now as the long, cold winter nights occlude our view of a proper path. As we continue to debate and deal with natural resources, primarily in economic terms, we seem to ask only how our decisions will affect us financially now and for the foreseeable future. We never get beyond the money. Our problems with the sea around us are many, and mostly of our own making, going back centuries. We have always had a complicated relationship with the sea. That relationship loses none of its intricacy as our understanding of the water and its inhabitants evolves, or of how our activities affect the water. And we continue to want it all.

We wonder if we can move into the spring confident that we can deal with problems in better harmony with the natural world and her rhythms than we have done before. We can no longer pay attention just to those pieces that affect us directly, taking them out of context on a species-by-species basis. We are beginning to think in terms of the ecosystem as a whole rather than as pieces of a giant jigsaw puzzle. Maybe, twelve generations later, we are finally beginning to learn what the Wampanoags tried to teach my own *Mayflower* ancestors so long ago and to help us resolve whether spring will bloom or if the winter darkness will prevail.

# Notes

The following references were particularly insightful or relevant for each of these chapters and are generally in the order the subject matter was discussed.

### Chapter 2 – Glaciers

*Oldale, Robert. *Cape Cod and the Islands, The Geologic Story.* Parnassus Imprints, East Orleans, MA, 208 pp. 1992.

*Strahler, Arthur. *A Geologist's View of Cape Cod.* Doubleday, NY, 115 pp. 1966.

> *These two books provide the basis for understanding the glacial processes that formed Cape Cod.

Giese et al., *Coastal Landforms and Processes at the Cape Cod National Seashore, Massachusetts – A Primer.* U.S. Geological Survey Circular 1417. 2015.

Dolin, Eric. *Leviathan.* W.W. Norton Co., 2007.

Kelly, Shawnie. *It Happened on Cape Cod.* A-Two-Dot Book. Morris Publishing, LLC. 2006.

Quinn, William. *Shipwrecks Around Cape Cod.* Parnassus. 1973.

Quinn. *Orleans: A Small Town with an Extraordinary History.* Lower Cape Publishing. 2012.

*Snow, Edward Rowe. *Storms and Shipwrecks of New England.* Yankee Publishing, 1946.

Tougias, Michael, and Sherman, Casey. *The Finest Hours.* Scribner, 2009.

> *The shipwreck books provide a perspective of storms at sea: the power of the ocean, the fear of the crew, the human drama that unfolds.

*Barnard, Ruth. *History of Early Orleans.* William S. Sullwold Publishing, 1975.

Deyo, Simeon. *History of Barnstable County, Massachusetts.* H.W. Blake and Co., 1890.

*Freeman, Frederick. *History of Cape Cod: The Annals of the Thirteen Towns in Barnstable County.* 1862.

> *The three history books provide a glimpse of how the towns developed based on the geology. Freeman and Deyo are the most comprehensive histories of Cape Cod.

## Chapter 3 – Conflicts seeking Resolution

> *Specific websites are mentioned here but not in the bibliography because websites or their addresses change frequently.

www.lobstermen.com/wp-content/uploads/2011/09/MASS-LOBSTER-INDUSTRY-2011.pdf.

www.mass.gov/eea/docs/dfg/dmf/publications/tr39-2006-lobster-report.pdf.

## Chapter 4 – Blubber Hunters

Braginton-Smith, John, and Oliver, Duncan. *Cape Cod Shore Whaling: America's First Whalemen.* History Press, Charleston, SC, 2008.

Linda Coombs (personal communication).

Earl Mills (personal communication).

*Dolin, Eric, *Leviathan.* W.W. Norton Co., 2007.

> Excellent resource on the history of whaling especially for the generally less familiar earlier days

Dow, George. "Whaleships and Whaling with an Account of the Whale Fishery in Colonial New England." Argosy Antiquarian, 1925.

*Dana, Richard. *Two Years Before the Mast.* Original printing 1869; reprint 2007.

> *Rich description of life at sea

Barkham, Michael. "A Survey of Basque Presence of Activity in Atlantic Canada in the Sixteenth Century." SSHRC Tourist Research Project, Sydney, NS. 2004.

*Haley, Nelson. *Whale Hunt.* Mystic Seaport Museum, 1990.

> *Illuminating firsthand account of whale voyage aboard *Charles W. Morgan.*

Songini, Marc. *The Lost Fleet.* St. Martin's Press, 2007.

Starbuck, Alexander. *History of the American Whale Fishery.* Castle Books, 1989.

Cumbler, John. *Cape Cod: An Environmental History of a Fragile Ecosystem.*

University of Massachusetts Press, 2014.

McGlynn, Daniel. "Whale Hunting: Should Whale and Dolphin Hunting be Outlawed?" CQ Researcher, 2012.

Sullivan, Robert. *Whale Hunt.* Simon and Shuster, 2000.

## Chapter 5 – Harpoons to Cameras

Center for Coastal Studies website

McGlynn, Daniel. "Whale Hunting: Should Whale and Dolphin Hunting be Outlawed?" 2012

Charles "Stormy" Mayo and Scott Landry (personal communications).

Abend, Alan, and Tim Smith. "Review of Distribution of the Long-finned Pilot Whale, *Globicephala melas,* in the North Atlantic and Mediterranean." NMFS NE-117, 2000.

http://www.rawstory.com/rs/2012/07/02/whaling-nations-defeat-proposed-atlantic-sanctuary/
http://www.pbs.org/wgbh/americanexperience/films/whaling/pdf

Mystic Seaport website, 38th voyage.

## Chapter 6 – High and Dry

Brian Sharpe, International Fund for Animal Welfare (personal communication).

https://storify.com/action4ifaw/ifaw-mmrr-team-responds-to-mass-dolphin-stranding Audubon.

Bob Prescott, Wellfleet Bay Wildlife Sanctuary (personal communication).

*Saving Sea Turtles: Preventing Extinction.* Documentary film by Michelle Gomes and Jennifer Ting.

http://www.capecodtimes.com/news/20170110/film-documents-record-turtle-strandings-on-cape-cod.

## Chapter 7 – Unintended Consequences

Palumbi, Stephen, and Carolyn Sotka. *The Death and Life of Monterey Bay, A Story of Revival.* Island Press/Shearwater Books, 2011.

Lelli, Barbara, and David E. Harris, "Seal Bounty and Seal Protection Laws in Maine: 1872-1972." Natural Resources Journal, 2006.

Lelli and Harris. "Seal and Bounty Protection Laws in Maine 1872-1972:

Historic Perspectives on a Current Controversy." Northeastern Naturalist, 2009.

http://www.smithsonianmag.com/science-nature/otters-the-picky-eaters-of-the-pacific-49760786/?no-ist

https://www.youtube.com/watch?v=uIZLY0BvkJc

US Congress Merchant Marine and Fisheries Committee Report 1971b: 11-12.

http://www.nefsc.noaa.gov/ecosys/ecology/ProtectedSpecies/Pinnipeds/

## Chapter 9 – Tragedy of the Commons

Hardin, Garrett. "Tragedy of the Commons." Science, 1968.

## Chapter 10 – Rise of King Cod

Fagan, Brian. *Fish on Friday*. Basic Books, 2006.

Kurlansky, Mark. *World Without Fish*. Workman Publishing, 2011.

Kurlansky. *Cod: A Biography of the Fish that Changed the World*. Penguin, 1997.

Kurlansky. *Salt: A World History*. Penguin. 2002.

Kurlansky. *Basque History of the World*. Walker and Co. 1999.

Deyo, Simeon. *History of Barnstable County, Massachusetts*. H. W. Blake and Co., 1890.

*Barkham, Michael. "A Survey of Basque Presence and Activity in Atlantic Canada in the Sixteenth Century." Prepared for the SSHRC Heritage Tourism Research Project, New Dawn Enterprises, Sydney, NS, 2004.

> *The Basques were important for both early whaling and fishing in the Northwest Atlantic.

*Bigelow, Henry B., and William C. Shroeder, *Fishes of the Gulf of Maine*. Fishery Bulletin 74, U.S. Department of the Interior, Fish and Wildlife Service, Washington, DC, 1953, 577 pp.

> *The essential reference for fishes in the Gulf of Maine specifically and North Atlantic generally.

Bolster, W. Jeffrey. *The Mortal Sea: Fishing the Atlantic in the Age of Sail*. Belknap Press of Harvard University Press, Cambridge, MA, 2012, 378 pp.

Bolster. "Opportunities in Marine Environmental History." Environmental History 11(2006):567-597.

Bolster. "Putting the Ocean in Atlantic History, Maritime Communities and Marine Ecology in the Northwest Atlantic 1500-1800." American Historical

Review, 19-47, February 2008.

Bolster. "Fishermen and Politicians: A Lost Alliance." Boston Globe, December 16, 2012
http://www.bostonglobe.com/ideas/2012/12/16/fishermen-and-politicians-lost-alliance/IndbdEcJwXIIY0AHcqCs1I/story.html

*Roberts, Callum. *An Unnatural History of the Sea.* Island Press/Shearwater Books, 2007.

> *The early fishing industry in the Northwest Atlantic is explained in detail through the work of Bolster and Roberts.

### Chapter 11 – Warm and Cold, Deep and Shallow

Ferguson, Mark. "Hard Racket for a Living: Making Light-Salted Fish on the East Coast of Newfoundland." Materiel Culture Review, Vol. 45, Spring, 1997.

Deyo, Simeon. *History of Barnstable County, Massachusetts.* H. W. Blake abnd Co., 1890.

Quinn, William. *The Saltworks of Historic Cape Cod.* Parnassus, 1993.

Brereton, John. *Discoverie of the North Part of Virginia.* 1602.

Winslow, Edward. (ed.) *Mourt's Relation: A Journal of the Pilgrims at Plymouth.* 1622.

McFarland, Raymond. *A History of New England Fisheries.* University of Pennsylvania. 1911.

Bolster, W. Jeffrey. *The Mortal Sea: Fishing the Atlantic in the Age of Sail.* Belknap Press of Harvard University Press, 2012.

Cumbler, John T. *Cape Cod: An Environmental History of a Fragile Ecosystem.* University of Massachusetts Press, 2014.

Bigelow, Henry, and William Schroeder. *Fishes of the Gulf of Maine.* 1953.

Freeman, Frederick. *History of Cape Cod: The Annals of the Thirteen Towns of Barnstable County.*1862.

Kittredge, Henry C. *Cape Cod, Its People and Their People.* Houghton Mifflin, 1930.

### Chapter 12 – Three Humble Fish

McKenzie, Matthew. *Clearing the Coastline: The Nineteenth-Century Ecological and Cultural Transformation of Cape Cod.* University Press of New England, 2010.

Hoy, Dwight, and George M. Clark, *Atlantic Mackerel Fishery 1804-1965.* Leaflet 603, Bureau of Commercial Fisheries, 1967.

Bolster, W. Jeffrey. *The Mortal Sea: Fishing the Atlantic in the Age of Sail.* Belknap Press of Harvard University Press, 2012.

McFarland, Raymond. *A History of New England Fisheries.* University of Pennsylvania, 1911.

Goode, George Brown. *A Short History of the Menhaden; An Abstract of A History of the Menhaden.* Salem Press, 1880.

Bigelow, Henry, and William Schroeder. *Fishes of the Gulf of Maine.* 1953.

## Chapter 13 – Herring

Hay, John. *The Run.* Beacon Press, 1999.

Whitborne, Charles. *The True Travels of Captain John Smith.*

Bigelow, Henry, and Schroeder, William. *Fishes of the Gulf of Maine.* 1953.

Cumbler, John T. *Cape Cod: An Environmental History of a Fragile Ecosystem.* University of Massachusetts Press, 2014.

## Chapter 14 – Growth to Devastation

Peluso, Charles "Tiggie," and Macfarlane, Sandy. *Tiggie: The Lure and Lore of Commercial Fishing in New England.* iUnverse, 2007.

Carey, Richard Adams. *Against the Tide: The Fate of the New England Fisherman.* Houghton Mifflin, 1999.

Warner, William W. *Distant Water: The Fate of North Atlantic Fishermen.* Little Brown and Co., 1977.

Dutra, Judy. *Nautical Twilight: The Story of a Cape Cod Fishing Family.* Create Space, 2011.

> *The following are general interest books about life on Cape Cod in the first half of the twentieth century.

Eldredge, Dana. *Cape Cod Lucky,* 2000; *Once Upon a Cape Cod.* Stony Brook Publishing, 1997.

*Kittredge, Henry C. *Cape Cod: Its People and Their People.* Houghton Mifflin, 1930.

Nickerson, Joshua. *Days to Remember.* Chatham Historical Society, 1988.

*Nickerson, William Sears. *The Bay as I See It.* Reprinted Friends of Pleasant Bay, 1995.

*Sparrow, Don. *Growing Up on Cape Cod.* Great Oak Publishers, 1999.

Sheedy, Jack, and Coogan, Jim. *Cape Cod Companion.* Harvest Home Books, 1999.

*Sheedy and Coogan. *Cape Cod Voyage.* Harvest Home Books, 2001.

## Chapter 15 – Invaded Territory and Clashing Schemes

Carey, Richard Adams. *Against the Tide, The Fate of the New England Fisherman.* Houghton Mifflin, 1999.

Warner, William W. *Distant Water: The Fate of North Atlantic Fishermen.* Little Brown Co, 1977.

Dutra, Judy. *Nautical Twilight: The Story of a Cape Cod Fishing Family.* Create Space, 2011.

McFarland, Raymond. *A History of New England Fisheries.* University of Pennsylvania, 1911.

Bolster, Jeffrey. *The Mortal Sea: Fishing the Atlantic in the Age of Sail.* Bellknap Press of Harvard University Press, 2012.

Sabine, Lorenzo. *Report of the Principal Fisheries of the American Seas,* Treasury of the United States, 1853.

Playfair, Susan R. *Vanishing Species: Saving the Fish, Sacrificing the Fisherman,* University Press of New England, 2003.

Dobbs, David. *The Great Gulf: Fishermen, Scientists and the Struggle to Revive the World's Greatest Fishery.* Island Press/Shearwater Books, 2000.

Bigelow, Henry, and Schroeder, William. *Fishes of the Gulf of Maine.* 1953.

Christy, Francis T., and Scott, Anthony. *The Common Wealth in Ocean Fisheries.* Johns Hopkins Press, 1965.

Bolster, Jeffrey. "Opportunities in Marine Environmental History." Environmental History 11:567-597, 2006.

Bolster. "Putting the Ocean in Atlantic History: Maritime Communities and Marine Ecology in the Northwest Atlantic 1500-1800." American Historical Review, 19-47, Spring, 2008.

Boston Globe, December 16, 2012.

Apollonio, Spencer. *Hierarchical Perspectives on Marine Complexities.* Columbia University Press, 2002.


## Chapter 16 – Estuaries and Changing Baselines

*Belding, David L. *The Works of David L. Belding, M.D., Biologist: Early 20th Century Shellfish Research in Massachusetts; Quahog and Oyster Fisher; The Scallop Fishery; The Soft-Shell Clam Fishery.* Republication, Cape Cod Cooperative Extension of Barnstable County, 2004.

   *Essential reference for biology and ecology of major commercial shellfish species.

Macfarlane, Sandy. *Rowing Forward, Looking Back: Shellfish and the Tides of Change at the Elbow of Cape Cod.* Friends of Pleasant Bay, 2002.

Massachusetts Division of Marine Fisheries Estuary Monographs: Pleasant Bay, 1967; Waquoit Bay, 1971; Wellfleet Harbor, 1972; Plymouth-Kingston-Duxbury, 1974; Plum Island Sound, 1968.

Massachusetts Estuary Program, Estuary Reports Pleasant Bay, and Nauset.

Roman et al. An ecological analysis of Nauset Marsh, Cape Cod National Seashore, 1989.

Wood, William; Alden T. Vaughan, Editor, *New England Prospect.* University of Massachusetts Press, 1977, 132 pp.

Geist, M.A., 1998. "Local and Global Effects of Human Induced Alterations to the Nitrogen Cycle: Waquoit Bay." NERR Science and Policy Bulletin Series No. 6, Waquoit Bay National Estuarine Research Reserve, Waquoit, MA.

*The following three titles are specific to barrier beach dynamics:

Aubrey, D.G, G. Voulgaris, W.D. Spencer, and S. O'Malley. "Tidal Residence Time Within the Nauset Marsh System." Report submitted to the Town of Orleans, Woods Hole Oceanographic Institution, Department of Geology and Geophysics, Woods Hole, MA, 1997.

Giese, G.S., D.A. Aubrey, and J.T. Liu. "Development, Characteristics and Effects of the New Chatham Harbor Inlet." Technical Report No. 89-19, Woods Hole Oceanographic Institution, Woods Hole, MA, 1989.

Speer, P.E. D.G. Aubrey, and E. Ruder. "Beach Changes at Nauset Inlet, Cape Cod, Massachusetts 1640-1981." Woods Hole Oceanographic Institution technical report, WHOI-82-40, Woods Hole, MA, 1982.

## Chapter 17 – Quahauging

Darling, Warren S. *Quahoging Out of Rock Harbor 1890-1930.* Privately printed by author, 1984.

> *The bullraking process was beautifully described in precise detail by Warren Darling in his booklet *Quohoging Out of Rock Harbor.* Bullraking continues today mostly in shallower waters than Cape Cod Bay but it is a relatively rare sight to see someone working with a bullrake compared to earlier times.

Belding, David L. "A Report Upon the Quahog and Oyster Fisheries of Massachusetts, Including the Life History, Growth and Cultivation of the Quahog," 1912.

## Chapter 18 – Mussels Above and Clams Below

Belding, David L. "The Soft-Shelled Clam Fishery of Massachusetts." 1907, republished in 1916 with new material and again in 1930.

## Chapter 19 – Nuggets of Gold

Belding. "The Scallop Fishery of Massachusetts Including an Account of the Natural History of the Common Scallop." 1910.

## Chapter 20 – Farming the Waters

Gef Flimlin, Sandy Macfarlane, Edwin Rhodes and Kathy Rhodes. "Best Management Practices for the East Coast Shellfish Aquaculture Industry." East Coast Shellfish Growers Association, NRAC and NOAA, 2010.

Dorothy Leonard and Sandra Macfarlane. "Best Management Practices for Shellfish Restoration." Interstate Shellfish Sanitation Conference, 2011.

## Chapter 21 – Shellfish to the Rescue

Massachusetts Division of Marine Fisheries Estuary Monographs: Pleasant Bay, Pleasant Bay, 1967; Waquoit Bay, 1971; Wellfleet Harbor, 1972; Plymouth-Kingston-Duxbury, 1974; Plum Island Sound. 1968.

Massachusetts Estuaries Project and TMDL Reports:
MA Estuaries Project Technical Report for Pleasant Bay.

Geist, M.A. *Local and Global Effects of Human Induced Alterations to the Nitrogen Cycle.* Waquoit Bay NERR Science and Policy Bulletin Series No. 6, Waquoit Bay National Estuarine Research Reserve, Waquoit, MA. 1998.

Reitsma, J., D. C. Murphy, A. F. Archer, and R. H. York, "Nitrogen Extraction Potential of Wild and Cultured Bivalves Harvested from Nearshore Waters of Cape Cod." Marine Pollution Bulletin, Vol. 116 (1-2) 2017,175-181. 2017.

## Chapter 22 – Ocean Conveyor

NOAA website is the most useful source of information and maps/graphic representations

> *The following technical papers cover specific topics within the general subject of the Gulf Stream

* Delworth, T. L., P. U. Clark, M. Holland, W. E. Johns, T. Kuhlbradt, J. Lynch-Stieglitz, C. Morrell, R. Seager, A. Weaver, and R. Zhang. "The Potential for Abrupt Change in the Atlantic Meridional Overturning Circulation,"

Chapter 4 in: *Abrupt Climate Change*. USG, 2008.

\* Drinkwater, K. F., A. Belgrano, A. Borja, A. Conversi, M. Edwards, C. H. Greene, G. Ottersen, A. J. Pershing, and H A. Walker. "The Response of Marine Ecosystems to Climate Variability Associated With The North Atlantic Oscillation." Chapter 10, J.W. Hurrell, et al. (ed.) *The North Atlantic Oscillation: Climate Significance and Environmental Impact*. Geophysical Monograph; 134, American Geophysical Union, Washington, DC, 2003.

\* Ezer, Tal and Larry P. Atkinson. "Accelerated Flooding Along the U.S. East Coast: On the impact of sea level rise, tides, storms, the Gulf Stream and the NAO." Earth's Future 2:362-382, 2014.

\* Frankignoul, Claude and Gaellede Coetlogo. "Gulf Stream Variability and Ocean and Atmospheric Interactions." Journal of Physical Oceanography 31(12):3516-3529, 2001.

Giese, G.S., S.J. Williams, and Mark Adams. *Coastal Landforms and Processes at the Cape Cod National Seashore, Massachusetts: A Primer*. U.S. Geological Survey Circular 1417, 86 p., 2015.

Hale, Stephen S. "Biogeographic Patterns of Marine Benthic Macroinverebrates Along the Atlantic Coast of Northeast U.S.A." Estuaries and Coasts 33:1039-1053, 2010.

\* Hurrell, James, Yochanar Kushnir, Geir Ottersen, Martin Visbeck, (eds.) "The North Atlantic Oscillation Climatic Significance and Environmental Impact." Geophysical Monograph Series 134, American Geophysical Union, 2003.

\* Rahmstorf, Stefan, J.E. Box, F. Fuelner, M.E. Mann, A. Robinson, S. Rutherford, and E.J. Schaffernicht. "Exceptional Twentieth Century Slowdown in Atlantic Ocean Overturning Circulation." Nature Climate Change 5:475-479, 2015.

## Chapter 23 – Challenges

Abel, David. "As Seas Rise City Mulls a Massive Sea Barrier Across Boston Harbor." Boston Globe, February 8, 2017.

Hale, Stephen. "Biogeographic Patterns of Marine Benthic Macroinverebrates Along the Atlantic Coast of Northeast U.S.A." 2010.

Pleasant Bay Resource Management Plan. 1998.

Final Management Plan Stellwagen Bank National Marine Sanctuary. 2010.

Demystifying Ecosystem-Based Fisheries Management. 2015. www.st.nmfs.noaa.gov/ecosystems/ebfm/ebfm-myths

Appendix
# Common and Scientific Names of Species Mentioned in the Text

**Mammals**
Harp seals (Pagophilus groenlandicus)
Hooded seals (Cystophora cristata)
Harbor seals (Phoca vitulina)
Gray seals (Halichoerus grypus)
Pacific Sea otter (Enhydra leutris)
Walrus (Odobenus rosmarus)
Bowhead whale (Balaena mysticetus)
Gray whale (Esochrichtilus robustus)
Blue whale (Belaenopters musculus)
Sperm whale (Physeter catadon)
Killer whale (Orcinus orca)
North Atlantic right whale (Eubalaena glacialis)
Humpback whale (Megaptera novaeangliae)
Minke whale (Balaenoptera acutorostrata)
Fin whale (balaenoptera physalus)
Pilot whales; Grampus; Blackfish (Globicephala melas – longfin); (Globicephala
    macrorhynchus – shortfin)
Bottle nose dolphin (Tursiops truncates)
Short beaked common dolphin (Delphinus delphis)

**Turtles**
Green turtle (Chelonia mydas)
Hawksbill turtle (Eretgmochelys imbricate)
Kemp's ridley turtle(Lepidochelys kempii)
Leatherback turtle (Dermochelys coriacea
Loggerhead turtle (Caretta caretta)

## Fish, Shellfish and Invertebrates

Sea herring (Clupea harrengus)
River herring – alewives (Alosa pseudoharengus)
River herring – bluebacks (Alosa aestivalis)
Winter flounder (pseudopleuronectes americanus)
Summer flounder – fluke (Paralichthys dentatus)
Sole (Glyptocephalus cynoblossus)
Yellowtail Flounder (Limanda ferruginea)
Squid (Loligo pealeii)
Skate (Leucoraja sp., Malacorajs sp., barndoor skate (Dipterus laevis)
Sculpin (Myoxocephalus sp.)
Cod (Gadus morhua)
Haddock (Melanogrammus aeglefinnus)
Ling (Ophidon elongates)
Pollock (Pollachius virens)
Halibut (Hippoglossus hippoglossus)
Cusk (Brosme brosme)
Atlantic tomcod (Microgadus tomcod)
Cunner (Tautoglabatus adspertus walbaum)
Tautog (Tautogis onitis)
Mackerel (scomber scombus)
Turbot (Scophthalmus maximus)
Mullet (Mugil cephalus)
Menhaden (Brevoortia tyrannus)
Butterfish (Peprilus triacanthus)
Scup (Stenotomus chrysops)
Black bass (Centropristis striata)
Sturgeon (Acipenser oxyrhinchus)
Striped bass (Morone saxtilis)
Bluefish (Pomotomus saltatrix)
Sand lance (Ammodytes americanus)
Bluefin tuna (Thunnus thynnus)
Yellowfin tuna (Thunnus albacares)
Hake – Red (Urophycis chuss); White (U. tenuis); Silver: whiting (Merluccius
    bilinearis)
Wolffish (Anarhicas lupus)
Sea robin (Prionotus carolinus)
Capelin (Mallotus villosus)
Eel (Anguilla rostrada)
Atlantic salmon (Salmo salar)
Alaska Pollock (Gadus chalcogrammus)

Pacific cod (Gadus microcephalus)
Pink salmon (Oncorhynchus gorbuscha)
Spiny dogfish (Squalus acantheas)
[Great] White shark (Carcharodan carcharias)
Lobster (Homarus americanus)
Blue Crab (Callinecte sapidus)
Horseshoe crab (Limulus polyphemus)
Clam, soft shell (Mya arenaria)
Quahaug, hard shell clam (Mercenaria mercenaria)
Razor clam (Ensis directus)
Bay scallop (Argopecten irradians irradians)
Blue mussel (Mytilus edulis)
American oyster (Crassostrea virginica)
Gem clam (Gemma gemma)
Knobbed whelk (Busycon carica)
Channeled whelk (Busycotypus canaliculatus)
Moon snail (Euspira heros)
Macoma (Macoma baltica)
Red tide, Alexandrium tamarenses
Beach grass (Ammophila brevigiligulata)
Marsh (low) grass (Spartina alterniflora)
Marsh (high) grass (Spartina patens)
Eelgrass (Zostera marina)
Sea lettuce (Ulva lactuca)
Green seaweeds (Entermorpha sp.),
Red/brown seaweeds (
Rockweed (Ascophyllum nodosum)

**Birds**
Canada geese, (Branta canadensis)
Brant (Branta bernicla)
Black duck (Anas rubripes)
Mallard (Anas platyrynchos)
Teal: blue winged (Anas discors); green winged (Anas crecca)
Widgeon (Anas Americana)
Canvasback (Aythya valisineria)
Eider (Somateria mollissima)
Bufflehead (Bucephala albeola)
Goldeneye (Bucephala islandica)
Scoter (Melanitta nigra)
Merganser (Mergus merganser)
Piping plover (Charadrius hiaticula)

Tern (Sterna hirundo)
Oyster catcher (Haematopus palliates)
Osprey (Pandion haliatus)
Marsh hawk, northern harrier (Circus cyaneus)
Black backed gull (Larus marinus)
Herring gull (Larus argentatus)
Laughing gull (Larus atricills)
White ibis (Eudocimus albus)

# Bibliography Presented in Approximate Sequence to the Book Sections

## Marine Mammals

Abend, Alan G. and Tim D. Smith. "Review of Distribution of the Long-finned Pilot Whale (Globicephala melas) in the North Atlantic and Mediterranean." NOAA Technical Memorandum NMFS-NE-117, 2000.

Apollonio, Spencer, Ed. *The Last of the Cape Horners: Firsthand Accounts from the Final Days of the Commercial Tall Ships,* Brassey's, Washington, D.C., 296 pp. 2000.

Bogomolni, A., G. Early, K. Matassa, O. Nichols and L. Sette. "Gulf of Maine Seals – Populations, Problems and Priorities." Woods Hole Oceanographic Institution Technical Report 2010-04, 2010.

Braiginton-Smith, John and Duncan Oliver, *Cape Cod Shore Whaling.* The History Press, Charleston, SC, 2008. 127 pp.

Carswell, Lilian. "Southern Sea Otter". US Fish and Wildlife Service, Ecological Services, California

Dana, Richard Henry, Jr. *Two Years Before the Mast,* Dover Publications, NY, 2007, 311 pp. 1869 (first publication).

Dolin, Eric Jay. *Leviathan.* W.W. Norton and Co., NY, 480 pp. 2007.

Dow, George Francis. *Whale Ships and Whaling with an Account of the Whale Fishery in Colonial New England,* 1925. Second Edition by Argosy Antiquarian, Ltd., New York, 446 pp. 1967.

Haley, Nelson Cole. *Whale Hunt.* Mystic Seaport Museum, Inc. Mystic,

CT, 304 pp. 1990.

LeCapra, Veronique. "Whale-Safe Fishing Gear: New Buoys for Lobster Traps Could Prevent Entrapments." Oceanus, November 1, 2016.

Lelli, Barbara and David E. Harris. "Seal Bounty and Seal Protection Laws in Maine: 1872-1972: Historic Perspectives on a Current Controversy." Natural Resources Journal 46(4):882-924. 2006.

Lelli, Barbara and David E. Harris. "Seal Bounties in Maine and Massachusetts: 1888-1962. Northeastern Naturalist 16(2):239-254. 2009.

McGlynn, Daniel. "Whale Hunting: Should Whale and Dolphin Hunting be Outlawed?" The CQ Researcher, Volume 22, Number 24. June 29, 2012.

Nichols, O. C., A. Bogomolni, E.C. Bradfield, G. Early, L. Sette, and S. Wood. "Gulf of Maine Seals: Fishery Interactions and Integrated Research." Provincetown Center for Coastal Studies and Marine Mammal Center, Woods Hole Oceanographic Institute. 2011.

Palumbi, Stephen R. and Carolyn Sotka. *The Death and Life of Monterey Bay, A Story of Revival.* Island Press/Shearwater Books, Washington, D.C. 211. pp.2011.

Songini, Marc. *The Lost Fleet.* St. Martin's Press, New York, 432 pp. 2007.

Spratt, Ashley. *"Slowly Swimming Towards Recovery, California's Sea Otter Numbers Holding Steady,"* US Fish and Wildlife Service. September 22, 2014.

Starbuck, Alexander. *History of the American Whale Fishery.* Castle Books, Seacaucus, NJ, 779 pp.1989.

Sullivan, Robert. *A Whale Hunt, How a Native American Village Did What No One Thought They Could,* Simon and Schuster, New York, 285 pp. 2000.

Tower, Walter Sheldon. *A History of the American Whale Fishery.* University of Philadelphia, 144 pp. 1907.

Trull, Peter. *The Gray Curtain: The Impact of Seals, Sharks and Commercial Fishing on the Northeast Coast.* Schiffer Publishing, Ltd. Atglen, PA, 79 pp. 2015.

Ward, Nathalie. *Stellwagen Bank: A Guide to the Whales, Sea Birds, and Marine Life of the Stellwagen Bank National Marine Sanctuary.* Downeast Books.

Camden, Maine. 232 pp. 1995.

US Congress Merchant Marine and Fisheries Committee Report 1971b: 11-12

**Fisheries**

Atlantic States Marine Fisheries Commission www.asmfc.org/about-us/program-overview.

Apollonio, Spencer. *Hierarchical Perspectives on Marine Complexities.* Columbia University Press, 229 pp. 2002.

Atlantic States Marine Fisheries Commission, Forging Knowledge Into Change: Commemorating 75 Years of Cooperative Sustainable Fisheries Management, 2016.

Barkham, Michael M. "A Survey of Basque Presence and Activity in Atlantic Canada in the Sixteenth Century. Prepared for the SSHRC Heritage Tourism Research Project, New Dawn Enterprises, Sydney, NS 2004.

Bigelow, Henry B. and William C. Shroeder. *Fishes of the Gulf of Maine,* Fishery Bulletin 74. U.S. Department of the Interior, Fish and Wildlife Service. Washingon, D.C. 577 pp. 1953.

Bolster, Jeffrey. *The Mortal Sea: Fishing the Atlantic in the Age of Sail.* Belknap Press of Harvard University Press. Cambridge, MA, 378 pp. 2012.

Bolster, Jeffrey. "Opportunities in marine environmental history," Environmental History 11:567-597. 2006.

Bolster, W. Jeffrey. "Putting the Ocean in Atlantic History, Maritime Communities and Marine Ecology in the Northwest Atlantic, 1500-1800." American Historical Review. 19-47. February, 2008.

Bolster, W. Jeffrey. "Fishermen and politicians: A lost Alliance, The collaboration That Emptied the New England Ocean Over Decades May Now be the Only Way to Bring Fish Back." Boston Globe, December 16, 2012.

Bourne, Russell. *The View from Front Street: Travels Through New England's Historic Fishing Communities.* W. W. Norton Co., NY, NY. 282 pp. 1989.

Carey, Richard Adams. *Against the Tide: the Fate of New England Fishermen.*

Houghton Mifflin Co. 381 pp. 1999.

Christy, Francis T. and Anthony Scott. *The Common Wealth in Ocean Fisheries.* Johns Hopkins Press. 281 pp. 1965.

Collette, Bruce B. and Grace Klein-MacPhee, (Eds.). *Bigelow and Shroeder's Fishes of the Gulf of Maine,* Third Edition, Smithsonian Institution Press, Washington, D.C., 748 pp. 2002.

Dobbs, David. *The Great Gulf: Fishermen, Scientists and the Struggle to Revive the World's Greatest Fishery.* Island Press/Shearwater Books. Washington. D.C. 2000.

Dutra, J. J. *Nautical Twilight: The Story of a Cape Cod Fishing Family.* Create Space, North Charleson, SC, 195 pp. 2011.

Fagin, Brian. *Fish on Friday.* Basic Books. New York. 338 pp. 2006.

Ferguson, Mark. "Hard Racket for a Living - Making Light-Salted Fish on the East Coast of Newfoundland," Materiel Culture Review, Volume 45. Spring, 1997.

Goode, G. Brown. *A Short History of the Menhaden; An Abstract of A History of the Menhaden,* Salem Press. 1880.

Hay, John. *The Run,* Beacon Press. Boston, MA. 164 pp. 1999.

Hoy, Dwight L. and George M. Clark, *Atlantic Mackerel Fishery 1804-1965.* Fishery Leaflet 603, U. S. Department of the Interior, Fish and Wildlife Service, Bureau of Commercial Fisheries, 1967, 9 pp.

Innis, Harold A. *The Cod Fisheries: the History of and International Economy.* University of Toronto Press, 522 pp. 1954.

Kulka, David W. History and Description of the International Commission for the Northwest Atlantic Fisheries www.nafo.int/Portals/0/PDFs/icnaf/ICNAF_history-kulka.pdf?ver=2016-08-10-090221-340

Kurlansky, Mark. *World Without Fish.* Workman Publishing Co., NY, NY. 184 pp. 2011.

Kurlansky, Mark. *Salt: A World History.* Penguin Books, NY, 484 pp. 2002.

Kurlansky, Mark. *Cod: A Biography of the Fish that Changed the World.* Penguin Books, NY. 294 pp. 1997.

Kurlansky, Mark. *Basque History of the World.* Walker & Co. NY, 387 pp. 1999.

Lear, W. H. "History of Fisheries in the Northwest Atlantic: The 500-year Perspective," J. Northw. Atl. Fish. CSci. 23:41-73. 1998.

LeHuenen, Joseph. "The Role of the Basque, Breton and Norman Cod Fishermen in the Discovery of North America from the XVIth to the End of the XVIIIth Century." Arctic, Vol. 37, No. 4 pp. 520-527. December 1984.

McFarland, Raymond. *A History of New England Fisheries.* University of Pennsylvania, 457 pp. 1911.

McKenzie, Matthew. *Clearing the Coastline: The Nineteenth-Century Ecological and Cultural Transformation of Cape Cod.* University Press of New England. Hanover, NH, 227 pp. 2010.

National Marine Fisheries Service Instruction 01-120-01 November 17, 2016. "Fisheries Management Ecosystem-Based Fisheries Management Policy NOAA Fisheries Ecosystem-Based Fisheries Management Road Map." 2016

Peluso, Charles "Tiggie" and Sandy Macfarlane. *Tiggie: The Lure and Lore of Commercial Fishing in New England.* iUniverse, 294 pp.2007.

Playfair, Susan R. *Vanishing Species; Saving the Fish Sacrificing the Fisherman.* University Press of New England. 270 pp. 2003.

Roberts, Callum. *An Unnatural History of the Sea.* Island Books/Shearwater Press, Washington,DC, 435 pp. 2007.

Sabine, Lorenzo. *Report of the Principal Fisheries of the American Seas.* Treasury of the United States, Washington, DC, 317 pp. 1853.

Shapiro, Sidney, Ed. *Our changing Fisheries,* U.S. Department of Commerce, 534 pp. 1971.

Serchuk, Frederic M. and Susan E. Wigley. "Assessment and Management of the Georges Bank Cod Fishery: An Historical Review and Evaluation." Journal of Northwest Atlantic Fishery Sciences Volume 13:25-52. 1992.

Warner, William, W. *Distant Water: The Fate of the North Atlantic Fisherman.* Little, Brown and Co., Boston, 338 pp. 1977.

Atlantic States Marine Fisheries Commission (ASMFC) website

International Commission for North Atlantic Fisheries (ICNAF) website

International Council for the Exploration of the Seas (ICES) website

ATMFC website

FAO website

A Snail's Odyssey website

United Nations Law of the Seas Convention Historical Perspective


**Environmental History/Estuaries**

Belding, David L. *The Works of David L. Belding, M.D., Biologist: Early 20th Century Shellfish Research in Massachusetts; Quahog and Oyster Fishery, The Scallop Fishery, The Soft-Shell Clam Fishery.* Republication of the works by Dr. David Belding. Cape Cod Cooperative Extension of Barnstable County. 2004.

Cramer, Deborah. *Great Waters,* W.W. Norton and Co., NY. 442 pp. 2001.

Cronon, William, ed. *Uncommon Ground: Rethinking the Human Place in Nature,* W. W. Norton and Sons, Co. NY. 561 pp.1996.

Cronon, William. *Changes in the Land: Indians, Colonists and the Ecology of New England.* Hill and Wang. NY. 241 pp. 1983.

Cumbler, John T. *Cape Cod: An Environmental History of a Fragile Ecosystem,* University of Massachusetts Press, Amherst and Boston, MA. 277 pp.2014.

Flimlin, Gef, Sandy Macfarlane, Edwin Rhodes and Kathy Rhodes. *Best Management Practices for the East Coast Shellfish Aquaculture Industry.* East Coast Shellfish Growers Association, NRAC and NOAA, 2010.

Geist, M.A. "Local and Global Effects of Human Induced Alterations to the Nitrogen Cycle" Waquoit Bay NERR Science and Policy Bulletin Series No. 6. Waquoit Bay National Estuarine Research Reserve, Waquoit, MA. 1998.

Giese, G.S., D.A. Aubrey and J.T. Liu. "Development, Characteristics and Effects of the New Chatham Harbor Inlet." Technical Report #89-19, Woods Hole Oceanographic Institution. Woods Hole, MA. 1989.

Leonard, Dorothy and Sandra Macfarlane. *Best Management Practices for Shellfish Restoration.* Interstate Shellfish Sanitation Conference, 2011.

Macfarlane, Sandy. *Rowing Forward, Looking Back: Shellfish and the Tides of Change at the Elbow of Cape Cod.* Friends of Pleasant Bay. Orleans, MA. 304 pp. 2002.

Macfarlane, Sandra L. "Bay Scallops in Massachusetts Waters: A Review of the Fishery and Prospects for Future Enhancement and Aquaculture." A Report to Barnstable County Cooperative Extension and Southeastern Massachusetts Aquaculture Center. 2004.

Macfarlane, Sandra L. "The evolution of a municipal quahaug (Hard Clam) Mercenaria mercenaria management program: a twenty year history -1975-1995." *Journal of Shellfish Research*, 17(4):1015-1036. 1998.

Macfarlane, Sandra L. "Shellfish as the impetus for embayment management." *Estuaries*, 19(2A): 311-319. 1995.

Macfarlane, Sandra L., "Shellfish enhancement programs: are they enough to maintain a fishery resource?" Office of Water Proceedings, 1994 Annual Meeting of the National Shellfisheries Association (Shellfish Stock Enhancement Session). USEPA 842-R-98-004, November, 1998.

Macfarlane, Sandra L. "Managing scallops Argopecten irradians irradians (Lamark, 1819) in Pleasant Bay, Massachusetts: large is not always legal." In Shumway, S.E. and P.A. Sandifer eds. *An International Compendium of Scallop Biology and Culture.* Baton Rouge, LA: World Aquaculture Society. pp. 264-272. 1991.

Massachusetts Division of Marine Fisheries Estuary Monograph Series. Pleasant Bay, 1967; Waquoit Bay, 1971; Wellfleet Harbor, 1972; Plymouth-Kingston-Duxbury, 1974; Plum Island Sound, 1968.

Massachusetts Estuaries Project and TMDL Reports: MA Estuaries Project Technical Report for Pleasant Bay

Reitsma, J, D.C. Murphy, A.F.Archer and R.H. York. "Nitrogen Extraction Potential of Wild and Cultured Bivalves Harvested from Nearshore Waters of Cape Cod," Marine Pollution Bulletin 116(1-2):175-181. 2017.

Roman, C.T., K.W. Able, K. I. heck, J. W. Portnoy, M.P. Fahay, D.G. Aubrey, and M.A. Lazzari. *An Ecological Analysis of Nauset Marsh, Cape Cod National Seashore*, National Park Service Cooperative Research Unit, Wellfleet, MA 1989.

Speer, P.E., D.G. Aubrey, and E.Ruder. "Beach Changes at Nauset Inlet, Cape Cod, Massachusetts 1640-1981." Woods Hole Oceanographic Institution technical report WHOI-82-40. Woods Hole, MA. 1982.

Valiela, I.K., K. Forman, M. LaMontagne, D. Hersh, J. Costa, P. Peckol, B. DeMeo-Anderson, C. D'Avanzo, M. Babione, C.H. Sham, J. Brawley, and K. Lajtha. "Coupling of Watersheds and Coastal Waters: Sources and Consequences of Nutrient Enrichment in Waquoit Bay, Massachusetts. Estuaries 15:443-457. 1992.

Valiela, I., G. Collins, J. Kremer, K. Lajtha, M. Geist, B. Seely, J. Brawley, and C.H. Sham. "Nitrogen Loading from Coastal Watersheds to Receiving Estuaries: New Method and Application." Ecological Applications 7(2):358-380. 1997.

Warren, Louis S., *American Environmental History*, Blackwell Publishers, Malden, MA. 2003.

## Geology/Oceanography/Climate

Aubrey, D.G., G. Voulgaris, W.D. Spencer and S. O'Malley. Tidal Residence Time Within the Nauset Marsh System. Report submitted to the Town of Orleans. Woods Hole Oceanographic Institution. Department of Geology and Geophysics. Woods Hole, MA. 1997.

Chamberlain, Barbara Blau. *These Fragile Outposts: A Geological Look at Cape Cod, Marthas Vineyard and Nantucket,* The Natural History Press. Garden City, NY, 327 pp. 1964.

Delworth, T.L., P.U. Clark, M. Holland, W.E. Johns, T. Kuhlbradt, J. Lynch-Stieglitz, C. Morrell, R. Seager, A. Weaver, and R. Zhang. *"The Potential for Abrupt Change in the Atlantic Meridional Overturning Circulation."* Chapter 4 in: Abrupt Climate Change, USGS. 2008.

Drinkwater, K. F., A. Belgrano, A. Borja, A. Conversi, M. Edwards, C. H. Greene, G. Ottersen, A. J. Pershing, and H A. Walker. "The Response Of Marine Ecosystems To Climate Variability Associated With The North Atlantic Oscillation." Chapter 10, JW Hurrell, et al. (ed.) *The North Atlantic Oscillation: Climate Significance and Environmental Impact.* Geophysical Monograph; 134, American Geophysical Union, Washington, DC, 2003.

Ezer, Tal and Larry P. Atkinson. "Accelerated Flooding Along the US

East Coast: On the Impact of Sea Level Rise, Tides, Storms, the Gulf Stream and the NAO," Earth's Future 2:362-382. 2014.

Frankignoul, Claude and Gaellede Coetlogon. "Gulf Stream Variability and Ocean and Atmospheric Interactions, Journal of Physical Oceanography." 31(12):3516-3529. 2001.

Giese, G.S., Williams, S.J., and Adams, Mark. *"Coastal Landforms and Processes at the Cape Cod National Seashore, Massachusetts—A Primer."* U.S. Geological Survey Circular 1417, 86 p. 2015.

Hale, Stephen S., "Biogeographic Patterns of Marine Benthic Macroinverebrates Along the Atlantic Coast of Northeast U.S.A.", Estuaries and Coasts 33:1039-1053. 2010.

Hurrell, James, Yochanar Kushnir, Geir Ottersen, Martin Visbeck, (eds.). "The North Atlantic Oscillation Climatic Significance and Environmental Impact," Geophysical Monograph Series 134, American Geophysical Union, 2003.

Maury, M.F. *The Physical Geography of the Sea.* Harper and Brothers, NY, 360 pp. 1858.

Mitchell, John Hanson. *Ceremonial Time: Fifteen Thousand Years on One Square Mile.* Anchor Press/Doubleday, NY. 222 pp. 1984.

Oldale, Robert N. *Cape Cod and the Islands, The Geologic Story,* Parnassus Imprints, East Orleans, MA, 208 pp. 1992.

Rahmstorf, Stefan, J.E. Box, F. Fuelner, M.E. Mann, A. Robinson, S. Rutherford, and E.J. Schaffernicht, "Exceptional Twentieth Century Slowdown in Atlantic Ocean Overturning Circulation," Nature Climate Change 5:475-479. 2015.

Strahler, Arthur N. *A Geologist's View of Cape Cod,* Doubleday, NY. 115 pp. 1966.

Tougias, Michael J. and Casey Sherman. *The Finest Hours,* Scribner, NY, 204 pp. 2009.

National Oceanic and Atmospheric Administration (NOAA) website

## Cape Cod/Regional

Barber, John Warner. *Historical Collections Being a General Collection of*

*Interesting Facts, Traditions, Biographical Sketches Anecdotes, etc., Relating to the History and Antiquities of Every Town in Massachusetts with Geographical Descriptions.* Dorr Howland and Co. Worcester, MA. *1841.*

Barbo, Theresa Mitchell. *Cape Cod Bay.* The History Press. Charleston, SC, 125 pp. 2008.

Barnard, Ruth L. *A History of Early Orleans.* William S. Sullwold Publishing. Taunton, MA. 158 pp. 1975.

Barnstable County. *Three Centuries of the Cape Cod County, Barnstable, Massachusetts, 1685-1985,* County of Barnstable. 437 pp. 1985.

Bradford, William. *Of Plymouth Plantation 1620-1647.* Alfred Knopf, Inc. NY. 448 pp.2002.

Brereton, John. *Discoverie of the North Part of Virginia,* 1602.

Brigham, Albert Perry. *Cape Cod and the Old Colony.* Grosset and Dunlap. NY. 284 pp. 1920.

Carr, Elliott. *Walking the Shores of Cape Cod.* On Cape Publications. 170 pp. 1997.

Coogan, Jim. *Sears Point,* Harvest Home Books, 233 pp. 2016.

Coogan, Jim and Jack Sheedy. *Cape Cod Harvest,* Harvest Home Books. East Dennis, MA. 208 pp. 2007.

Coogan, Jim. *Sail Away Ladies.* Harvest Home Books. Dennis, MA, 207 pp. 2003.

Coogan, Jim and Jack Sheedy. *Cape Cod Voyage, A Journey Through Cape Cod's History and Lore.* Harvest Home Books, 208 pp. 2001.

County of Barnstable. *Three Centuries of the Cape Cod County, Barnstable, MA, 1685-1985.* Barnstable County. 437 pp. 1985.

Dalton, J.W., *The Lifesavers of Cape Cod,* reprinted 1967 by Chatham Press, this edition Parnassus Imprints, Orleans, MA. 156 pp. 1902.

Darling, Warren S. Quahoging out of Rock Harbor 1890-1930. Privately printed by the author at Thompson's Printing, Inc. Orleans, MA. 1984.

De Champlain, Samuel. *Voyages of Samuel de Champlain 1604-1618,* Internet Archive

Deyo, Simeon L. (ed.) *History of Barnstable County, Massachusetts.* H.W. Blake & Co., New York, 1010 pp. 1890.

Digges, Jeremiah. *Cape Cod Pilot.* Modern Pilgrim Press. Provincetown, MA 403 pp. 1937.

Dwight, Timothy. *Travels in New England and New York.* 512 pp. 1823.

Eldridge, Dana. *Cape Cod Lucky.* Stony Brook Publishing and Productions, Inc., Brewster, MA, 2000, 170 pp.

Eldridge, Dana. *Once Upon Cape Cod,* Stony Brook Publishing and Productions, Inc. Brewster, MA. 112 pp. 1997.

Freeman, Frederick. *History of Cape Cod: The Annals of the Thirteen Towns of Barnstable County,* 1862.

Hawes, James W. "Library of Cape Cod History and Genealogy: Historical Address Delivered by James E, Hawes at the Occasion of the Celebration of the 200th Anniversary of Incorporation of Chatham, Confined Chiefly to the Period before 1860." C.W. Swift, Publisher and Printer. Yarmouthport, MA. The "Register" Press, 38 pp. 1912.

Jalbert, Russell R. *Where Sea and History Meet: 400 Years of Life in Orleans,* Orleans Bicentennial Commission Orleans, MA, 106 pp. 1997.

Kelley, Shawnie M. *It happened on Cape Cod,* A-Two-dot Book. Morris Publishing, LLC. 123 pp. 2006.

King, H. Roger. *Cape Cod and Plymouth Colony in the Seventeenth Century.* University Press of America. Lanham, MD, 307 pp. 1994.

Kittredge, Henry C. *Cape Cod, Its People and Their People,* Houghton Mifflin Co., Boston, MA. 330 pp. 1930.

Kittredge, Henry C. *Shipmasters of Cape Cod,* Parnassus Imprints. Hyannis, MA. 320 pp. 1963.

Nickerson, Joshua Atkins 2nd. *Days to Remember: A Chatham Native Recalls Life on Cape Cod Since the Turn of the Century.* The Chatham Historical Society. Chatham, MA, 277 pp. 1988.

Nickerson, W. Sears. *The Bay – As I See It;* Jack Viall, 47 pp. 1981.

Palfrey, John Gorham. *A Discourse Pronounced at Barnstable on the Third of September 1839 at the Celebration of the Second Centennial Anniversary of the*

*Settlement of Cape Cod.* Boston: Ferdinand Andrews. 1840.

Quinn, William P. *Orleans: A Small Cape Cod Town with an Extraordinary History,* Lower Cape Publishing. Orleans, MA. 267 pp. 2012.

Quinn, William P. *The Saltworks of Historic Cape Cod.* Parnassus Imprints. Orleans, MA, 247 pp. 1993.

Quinn, William P. *Shipwrecks Around Cape Cod.* Lower Cape Publishing Co. Orleans, MA. 240 pp. 1973.

Sheedy, Jack and Jim Coogan, *Cape Cod Companion,* Harvest Home Books. East Dennis, MA, 208 pp. 1999.

Small, Isaac M. *Shipwrecks of Cape Cod.* reprinted 1967 by Chatham Press, Chatham, MA, 86 pp. 1928.

Smith, Nancy W. Paine. *The Provincetown Book.* Tolman Print, Inc. Brocton, MA. 260 pp. 1922.

Snow, Edward Rowe. *Storms and Shipwrecks Around New England,* Yankee Publishing, Manchester, NH. 1946.

Sparrow, Donald. *Growing Up on Cape Cod: Four Brothers Learning to Stand Tall.* Great Oak Publishers, Eastham, MA 163 pp. 1999.

Thoreau, Henry David. *Cape Cod, 1817-1862.* Princeton University Press. Princeton, NJ. 235 pp. 1988.

Winslow, Edward, (ed.) Mourt's Relation: A Journal of the Pilgrims at Plymouth, 1622.

Wood, William. Alden T. Vaughan, (ed). *New England Prospect,* University of Massachusetts Press, 132 pp. 1977.

## Native American

McGaa, Ed, Eagle Man. *Nature's Way, Native Wisdom for Living in Balance with the Earth,* Harper Collins Publishers, NY, 285 pp. 2004.

Mills, Earl, Sr., Chief Flying Eagle and Betty Breen. *Cape Cod Wampanoag Cookbook: Wampanoag Indian Recipes, Images and Lore,* Clear Light Publishers, Santa Fe, NM, 190 pp. 2001.

Mills, Earl Sr. and Alicia Mann. *Son of Mashpee: Reflections of Chief Flying Eagle, A Wampanoag.* Word Studio, North Falmouth, MA, 115 pp. 1996.

Russell, Howard S. *Indian New England Before the Mayflower.* University Press of New England, Hanover, NH, 284 pp. 1980.

## General

Angus, Julie. *Rowboat in a Hurricane.* Greystone Books. Vancouver, BC, 267 pp. 2008.

Fontenoy, Maud. *Challenging the Pacific.* Arcade Publishing, Inc. NY. 155 pp 2006.

Outen, Sarah. *Dare to Do, Taking on the Planet by Boat and Bike,* Nicholas Bready Publishers. London. 289 pp. 2017.

Pope Nicholas V, Papal Bull, Dum Diversas, June 18, 1452; also Papal Bull Romanus Pontifex, January 5, 1455.

Ridgway, John and Chay Blyth. A Fighting Chance: How We Rowed Across the Atlantic in 92 Days. J. B. Lippincott Co. Philadelphia. 242 pp. 1967.

Plimouth Plantation website.

# About the Author

Sandy Macfarlane has lived her entire life by the sea. Like many kids, she grew up loving the water and her curiosity about all things marine, begun in youth, continues. Aided by forty years of hands-on experience in resource evaluation, policy and management and decades of recreational rowing in a fixed-seat skiff, she learned about the biological and physical world around her. In the process, she also gained insight into the social, political and economic landscape of coastal communities. She is past-president of the New England Estuarine Research Society and a member of numerous professional societies and community outreach organizations. *Swirling Currents* is her third book.

Additional Books by the author:
*Rowing Forward, Looking Back: Shellfish and the Tides of Change at the Elbow of Cape Cod*
*Tiggie: The Lure and Lore of Commercial Fishing in New England,* a collaboration with commercial fisherman Charles "Tiggie" Peluso, Independent Book Publishers Bronze Medal for Best Regional Non-fiction in 2009.